# CONTENTS

| | |
|---|---|
| Introduction | 3 |
| Highland Railway 4-6-0s | 4-9 |
| Growing up in Coventry | 10-25 |
| Derbyshire in colour | 26-27 |
| Midland to the core: James William Watkins | 28-30 |
| Garsdale | 31-37 |
| Recreating a colourful past | 38-41 |
| LMS General Instructions and Diagrams | 42-45 |
| LMS Accident Report: Plaistow | 46-47 |
| North by North West | 48-55 |
| Fury | 56-59 |
| Reminiscences of the LMS and MacBraynes | 60-65 |
| The 'Knotty' in the late steam age | 66-76 |
| The Platform End | 77-79 |
| Duchess of Gloucester | 80 |

Introduced to the Caledonian Railway in 1883, Drummond 'Standard Goods' Class 2F 0-6-0 No. 57339 is pictured near Killin Junction in 1959.
PHOTO: NEVILLE STEAD COLLECTION (NS204282) © THE TRANSPORT TREASURY

© Images and design: The Transport Treasury 2023. Design and Text: Peter Sikes
ISBN: 978-1-913251-61-1
First published in 2023 by Transport Treasury Publishing Ltd., 16 Highworth Close, High Wycombe HP13 7PJ

The copyright holders hereby give notice that all rights to this work are reserved.
Aside from brief passages for the purpose of review, no part of this work may be reproduced, copied by electronic or other means,
or otherwise stored in any information storage and retrieval system without written permission from the Publisher.
This includes the illustrations herein which shall remain the copyright of the copyright holder.

Copies of many of the images in MIDLAND TIMES are available for purchase/download.
In addition the Transport Treasury Archive contains tens of thousands of other UK, Irish and some European railway photographs.

www.ttpublishing.co.uk or for editorial issues and contributions email **MidlandTimes1884@gmail.com**

Printed in England by Short Run Press Limited, Exeter.

# INTRODUCTION

I'd like to start this third issue of Midland Times by thanking the contributors for the articles they send to me for inclusion in this publication. The quality of the articles is excellent, and if I thought that my railway knowledge was half-decent before I started putting Midland Times together then I have definitely been proved wrong.

In this issue a large part of Great Britain is covered. We start our journey in Scotland with a feature on Highland Railway 4-6-0 locomotives and follow this with an interesting and informative trip to Coventry. Initially services travelling south to London by-passed the city but as it grew in importance between the wars, more regular and faster trains started calling there en-route from and to Birmingham.

We then dart up and down the country visiting Derbyshire with two stunning colour photographs and then south to Ashchurch on the Birmingham to Bristol route. Garsdale is next up, a place which still provides us with stunning views today. There is a selection of images that show the amount of infrastructure built by the Midland Railway at this outpost deep in the fells of south east Cumbria.

Moving away from locations, David P. Williams explains how he produces his coloured monochrome images, three of which have adorned the front covers of Midland Times. This is followed by selected pages that have been reproduced from an original LMS booklet instructing staff the correct way to secure 'long and projecting traffic'.

Back on the rails we visit the scene of an accident at Plaistow then move back up the country to witness the 'last knockings' of steam in the north west accompanied by some very evocative photography.

*Fury* then takes over with an compelling article about the ill-fated high-pressure experiment pursued by the LMS. Another trip to Scotland follows which includes ferry trips with a company that would become part-owned by the LMS.

And, finally, there is a nostalgic trip to the 'Knotty' in 1962 showing some delightful scenes of the former North Staffordshire Railway network. A few miles covered there and hopefully a wide selection of subjects to interest you. Enjoy.

We are happy to announce that Midland Times has been well received and as a consequence will be increasing from two issues per year to three, and that the publication will be available on our subscription service.

Thank you for your support.

PETER SIKES, EDITOR, MIDLAND TIMES
email: midlandtimes1884@gmail.com

---

FRONT COVER (AND INSET RIGHT):

A Perth-Aberdeen service is seen at Platform 8 (one of the two north facing bays) at Perth General station in 1953 with No. 40921 at the head of the train. These semi-fast services to Aberdeen typically consisted of three coaches and a long-wheelbase van as seen here.

LMS-built 'Compound' numbers 40921-40924 were all allocated to Perth at this time with No. 40921 being the final survivor when withdrawn in November 1955.

There's plenty of human interest in this delightful scene. In the far distance a Dundee train can be seen passing at the end of the station.

© DAVID P. WILLIAMS COLOUR ARCHIVE.

# HIGHLAND RAILWAY 4-6-0s
## by Neville Stead

The 4-6-0 tender engine was the most prolific passenger locomotive of the twentieth century. Stanier's immortal 'Black Five' on the LMS ultimately totalled over 800 examples, while the LNER Thompson B1 topped 400. On the GWR, 330 Halls were built in addition to the numerous Castles, Granges, Stars, Saints and Manors.

Curiously, the first British 4-6-0 was built in 1894 for one of the smaller pre-grouping companies, the Highland Railway. The CME, David Jones, designed a class of 15 locos several years before the major railway companies, at a time when 4-4-0s were still the mainstay of passenger services, and four years before the first British 'Atlantic' appeared. When the 1923 Grouping took place, the Highland Railway had fifty 4-6-0s in service, of four different designs, out of a total of 173 locomotives handed over to the LMS – and when the oldest 2-4-0s and 4-4-0s, which never received (or were allocated) LMS numbers are discounted, the operational total drops to 150. So, one in three locomotives in service with the Highland Railway was a 4-6-0, the highest proportion of any British railway. There were a further six Highland-built 4-6-0s that were banned from the system! (See upper photo, page 7.)

Jones Goods No. 17930, ex-Highland Railway No. 117, this was the last of 15 to be built. The first, HR 103 (LMS No. 17916), was withdrawn in 1934 and is now part of the National Collection. Last to survive was No. 17925 which was withdrawn in 1940. No. 17930 is seen at Kyle of Lochalsh, circa. 1936 where the 'Jones Goods' were the regular locos for many years until the arrival of Stanier's Black Fives.
PHOTO: NEVILLE STEAD COLLECTION (NS204436) © THE TRANSPORT TREASURY

Peter Drummond, who succeeded Jones as CME in 1896, produced the 'Castle' class in 1900. Sixteen were produced initially with 5ft 9in driving wheels, and worked the passenger services between Perth, Inverness and Wick. A further three, with 6ft 0in wheels, were built in 1917 as replacements for the ill-fated 'Rivers'. No. 14685 DUNVEGAN CASTLE (HR No. 30) is seen at Dingwall. Withdrawals began in 1930, the last, No. 14690 DALCROSS CASTLE, being taken out of service in April 1947.
PHOTO: NEVILLE STEAD COLLECTION (NS204428) © THE TRANSPORT TREASURY

| HIGHLAND RAILWAY 4-6-0 LOCOMOTIVES – Specifications | | | |
|---|---|---|---|
| | CASTLE CLASS | CLAN CLASS | JONES GOODS |
| Designer | Peter Drummond | Christopher Cumming | David Jones |
| Builder | Dübs & Co./NBL, Glasgow | Hawthorn Leslie | Sharp, Stewart & Co. |
| Build date | 1900 | 1919 | 1894 |
| Total produced | 19 | 8 | 15 |
| Driving wheels | 5ft 9in (6ft 0in last three built) | 6ft 0in | 5ft 3½in |
| Bogie wheels | 3ft 3in | 3ft 0in | 3ft 2½in |
| Wheelbase | 26ft 3in | 25¾ft | 25ft |
| Loco weight | 58 tons 17cwt | 62 tons 4 cwt | 56 tons |
| Boiler | – | 4ft 9³⁄₁₆in diameter | 4ft 7⅞in diameter |
| Boiler pressure | 180psi | 175psi | 175psi |
| Tractive effort | 21,922 lbs | 23,012 lbs | 24,555 lbs |
| Coal capacity | 5 tons | 7 tons | – |
| Water capacity | 3,350 gallons (4,000 last three built) | 3,500 gallons | 3,000 gallons |
| Cylinders | 19½" x 26" | 21" x 26" | 20" x 26" |
| Valve gear | Allan Link | Walschaerts | Stephenson (slide valves) |

By 1917, F. G. Smith had come and gone and C. Cumming was in charge. A class of eight mixed-traffic 4-6-0s, with 5ft 3in driving wheels (as used on the Jones Goods), was put into traffic, again with the Kyle line as an obvious location, though it was only in their later days, after the Jones Goods were scrapped, that the 'Clan Goods' became a familiar sight on Inverness-Kyle trains. Numbered 75 to 82, (LMS 17950-17957) six survived into BR days, the last one, No. 57954, being withdrawn from Inverness in 1952. No. 17956 is illustrated on Inverness shed turntable on 6th June 1948. PHOTO: NEVILLE STEAD COLLECTION (NS204437) © THE TRANSPORT TREASURY

Two years after the previous class, Cumming produced another eight 4-6-0s, this time with 6ft 0in driving wheels, but very similar in appearance to his earlier locomotive designs. These were the 'Clan' class, which were the post-war front line locomotives for the principal Highland Railway passenger trains. Until the advent of LMS 'Black Fives' they were the mainstay of services north of Perth. In the late 1930s, superseded from their duties, they appeared on Glasgow-Oban trains for a while. Two survived into BR service – one, No. 54767 CLAN MACKINNON, ran with a BR number before withdrawal from Inverness in 1950, where it is seen by the 'triumphal arch', the distinctive feature of Inverness shed. PHOTO: NEVILLE STEAD COLLECTION (NS204438) © THE TRANSPORT TREASURY

LMS No. 14756, seen in the yard of the Highland Railway shed at Perth North. First of a class of six intended for use on the Highland Railway, it was designed by the CME, F. G. Smith, in 1915. This locomotive was built as HR No. 70 RIVER NESS, but the class was banned by the company's civil engineer as being too heavy for the railway infrastructure, and never worked on the system in pre-grouping days. They were sold immediately to the Caledonian Railway, and Smith found himself out of a job. Ironically in LMS days, after improvements to both track and bridges, the 'Rivers' made a successful reappearance on Perth-Inverness trains. The last two finished their days at Ayr, reprieved due to the Second World War, where they received the nicknames 'Gneisenau' and 'Scharnhorst'. The latter, No. 14760, was finally withdrawn in December 1946. PHOTO: NEVILLE STEAD COLLECTION (NS204209) © THE TRANSPORT TREASURY

'River' Class 4-6-0 No. 14690 pilots an unidentified Stanier 'Black Five' on an express working through Pitlochry in 1935.
PHOTO: NEVILLE STEAD COLLECTION (NS204429) © THE TRANSPORT TREASURY

# SCOTTISH RAIL TOUR
Stephenson Locomotive Society/Railway Correspondence & Travel Society
12th – 17th June 1960

15th June 1960 • Preserved Highland Railway 'Jones Goods' Class 4-6-0 No. 103 pictured at Inverness.
This was the only leg of the tour that no. 103 would participate in, the route being as follows:
Inverness – Tomatin – Aviemore – Kingussie – Dalwhinnie – Blair Atholl – Pitlochry – Stanley Jnc – Almond Valley Jnc – Perth
Photo: John McCann (OTA2407241) © The Transport Treasury

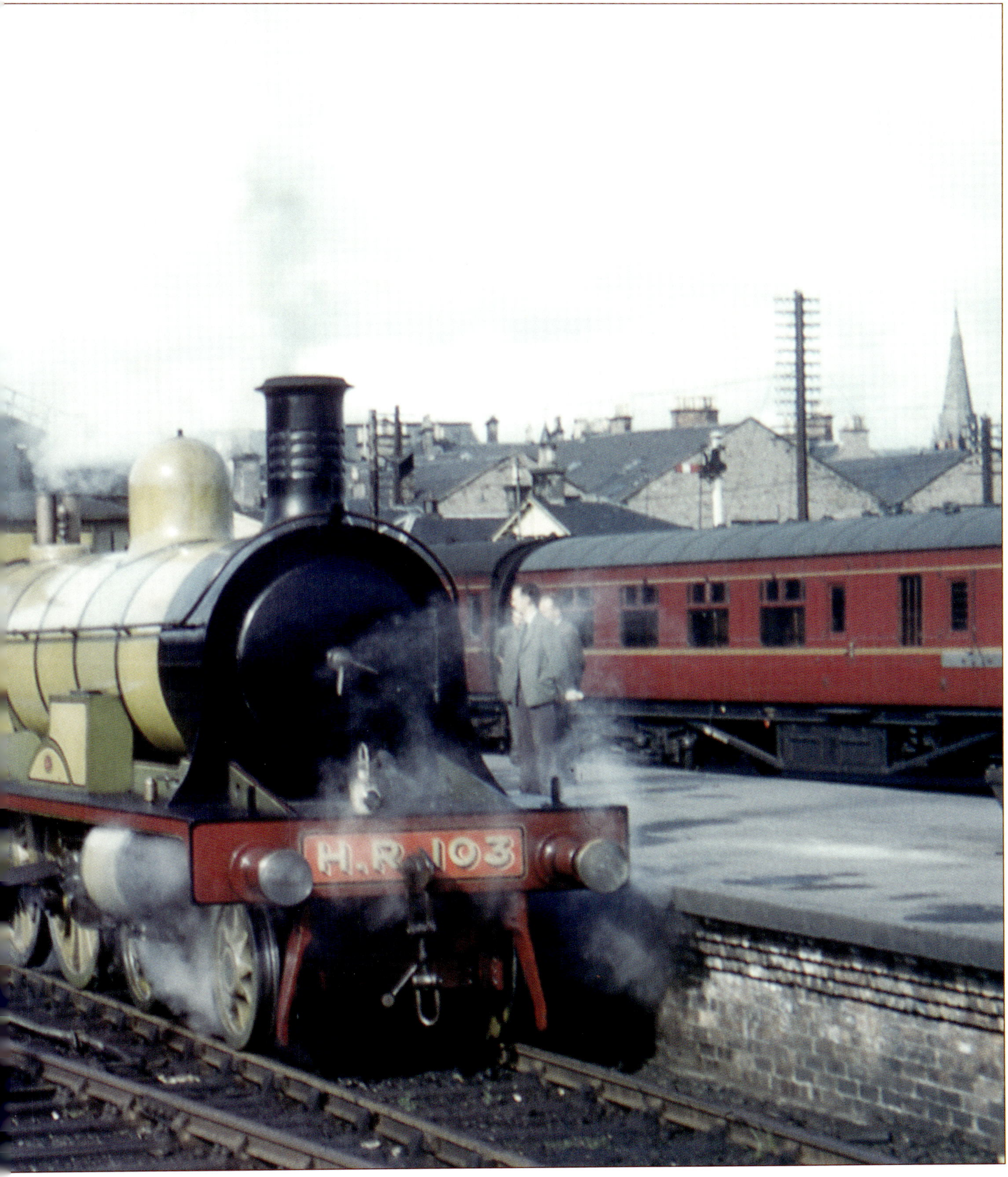
H.R 103

## GROWING UP IN COVENTRY

### by Ian Lamb

My association with Coventry began during the Second World War. Glasgow Clydebank was almost flattened by the 'blitz' from German aircraft during May 1941. I was born shortly after at the other side of Scotland in Restalrig, Edinburgh. Margaret Glass (my elder cousin) was my mother's favourite niece, so she was chosen to be my Godmother, and presented me for baptism in the Canongate Kirk, the Church of Holyrood House, within my father's parish of Abbeyhill.

Shortly after my birth, Margaret was posted for 'war duty' constructing munitions at the Daimler factory in Coventry, during which time she met – and married – Ted Souter – a miner from the north east of England, who had been transferred to Coventry Colliery. Initially the Souters were housed in a farm house at Exhall, where my abiding memories are of picking damsons, and my 'fear' of using the outside 'dry' toilet where the seat appeared very large, but being a small child was concerned that I might fall through the hole!

The main recollection of that first visit to Coventry was immediately 'post-war' when my father had chosen to take the very long route from Edinburgh to Coventry via London Kings Cross, Marylebone and Rugby, staying overnight in London with family friends, compared to most of my journeys in later childhood direct from Edinburgh Princes Street (via the west coast) to Birmingham, and on to Coventry.

My father's action was simply one of cost. Prior to 1948 (when the railways were nationalised) he was solely employed by the LNER, so his 'free' pass could only be used to Rugby Central, leaving the short link between Rugby Midland and Coventry on LMS tracks needing to be paid for.

During 1950 the Souters moved into miner's housing in the community of Keresley End on the outskirts of the city, which in many ways became my second home throughout my youthful years. Being close to the Coventry 'Pit' gave me opportunity to spend time at Ivy Cottage, Wheelright Lane, near Rowley's Green level crossing watching colliery trains heading to-and-fro between the mine and the through railway line linking Coventry to Nuneaton.

Of course, the main attraction was Coventry station which was visited whenever the No. 16 bus was taken into the city centre. In these days only two platforms were accessed as loops from the main through tracks. On one of my first 'spotting' forays while

Coventry station in early British Railways days, then part of the London Midland Region. The rather impressive curved canopy is covering a great variety of taxis and period cars.
PHOTO: MILEPOST 92½ (MC10058CA) © THE TRANSPORT TREASURY

30TH APRIL 1963 • Ex-WR Hawksworth designed 1949 Pannier tank locomotive No. 1509 (left) was a strange 'bed-fellow' in finding itself based at Coventry Colliery. Its duties were also shared on 5TH MARCH 1967 with No. 1501 (right) of the same 1500 class. At one time this latter engine had its wheel connecting rods, tank and cab sides painted dark red. Note the pit-head equipment in the background.
PHOTOS © THE TRANSPORT TREASURY – LEFT: DICK RILEY (RCR17003). RIGHT: ALEC SWAIN (AS V43-2)

JULY 1953 • Introduced in 1890 by the LNWR, Class 1P No. 46654 rests in Coventry station. (I wonder if that's me standing in front of the engine as it was about to 'blow off'?). PHOTO: ALEC FORD (AF0551) © THE TRANSPORT TREASURY

standing near the Birmingham end of the 'Down' platform, I was busy concentrating on the intricacies of a shunting locomotive (can't remember which one) when it 'blew off'. The sound of escaping steam pressure remains with me today, and scared me to such an extent that I've never ventured on to the footplate of a 'live' engine ever since, apart from a preserved railway locomotive.

I'm a great believer in the value of named trains, especially in the way of promotion and romanticism of specific trains. *The Flying Scotsman* and *The Royal Scot* are two perfect examples that have stood the test of time, yet today's trains simply exist as a means of getting the traveller from one place to another, rather than a specific chosen journey to enjoy and feel special.

Prior to the complete rebuilding of the Coventry station complex, these two views show the openness of the original layout when only two platforms existed. The top picture is looking from the main entrance platform towards Birmingham, whilst the lower image predominantly focuses on the routes to Rugby (straight ahead) and Leamington via Kenilworth (sharp right) past the engine shed. PHOTOS: MILEPOST 92½ (MC10014CA/CB) © THE TRANSPORT TREASURY

A motley collection of original LMS coaches in 'blood and custard' livery behind a Hughes 'Crab' 2-6-0 locomotive as it waits for clearance to continue its journey westwards. Meanwhile a 2P Class 4-4-0 No. 40678, based at Stafford (5C), proceeds slowly 'light engine' on the 'Up' main line to be halted at the 'home' lower quadrant signal by the steel overbridge, for authority to proceed ahead, probably for servicing at the engine shed. The advertisements (right) display the messages of the time, whilst a lone passenger strolls along to the end of the 'Up' platform. PHOTOS: MILEPOST 92½ (MI0014CC/CD) © THE TRANSPORT TREASURY

24TH OCTOBER 1959 • A class 101 DMU prepares to head off to Northampton. PHOTO: DICK RILEY (RCR 14407) © THE TRANSPORT TREASURY

24TH OCTOBER 1959 • Second of the Stanier 'Jubilee' Class 4-6-0 No. 45553 CANADA, based at Crewe, rests at Coventry station before heading westwards with its 'Up' train towards Birmingham. Apart from the new housing in the background, the start of the renovation of the old platform is under way. PHOTO: DICK RILEY (RCR 14411) © THE TRANSPORT TREASURY

*The Jacobite* tourist train between Fort William and Mallaig is almost always packed for its daily journey, yet the local Class 156 two-coach DMU is rarely full, even though its large windows provide better views of the spectacular surrounding scenery... and a cheaper fare! What the latter lacks – like any unnamed train – is aura and atmosphere.

The great Cecil J. Allen reminds us that "as far back as 1902, the London & North Western Railway first timed a non-stop train to cover the 112.9 miles between London and Birmingham in the even two hours; this was the 5pm from New Street to Euston. The route was ideal for the purpose; apart from the initial climb of 1½ miles from the platform end at Euston to Camden, at 1 in 70 to 1 in 105, and the short and sharp rise into New Street, Birmingham, the old London & Birmingham Railway was so magnificently engineered that there was no other gradient steeper than 1 in 330 throughout its length. Moreover, the only speed restriction enforced at any point was the 40mph through Rugby."

In the early years, Coventry had not yet established itself with motor manufacturing eminence, but prior to the First World War two through trains from Euston to Birmingham (the 8.40am and 6.55pm) serviced Coventry in the way of 'slip coaches', in order to maintain the two-hour schedule. These coaches were stopped clear of the platform loop, and then drawn into the station by waiting locomotives or horses. In 1914, the LNWR specially built vestibule slip coaches to give Coventry passengers access to the restaurant cars up to the moment of severing the train. Although the slip coaches were not restored after 1919, the importance of the City of Coventry was increasing rapidly so an adequate service had to be provided. More powerful locomotives became available so with faster running and increased train weight, an intermediate stop at Coventry could be incorporated within the two-hour schedule between Birmingham and Euston.

During the 1950s, the City of Coventry was bypassed by the great named trains out of Euston to the north, except *The Midlander*, which of itself was one of five West Coast expresses beginning with the letter 'M'. During this period the London Midland Region strove hard to create a corporate identity for itself in the early months of Nationalisation.

Initially, on 25th September 1950, the title was placed on the 9.45am Wolverhampton to London (via the Northampton loop), and the 5.45pm return journey. Three years later, travel times were changed, with *The Midlander* becoming an 11am departure from Wolverhampton, followed with a return journey north from Euston at 5.30pm. This timetable continued until 11th September 1959 when the title was withdrawn.

During any train-spotting excursion, a visit to the local motive power depot was always an attraction, and such a call at Coventry engine shed (2B) was no exception. Like most steam environments in the 1960s, such locomotives were being stored or withdrawn, mainly in favour of diesel, and especially electric. During this period the Coventry station complex was being fully rebuilt, particularly with four platforms in place of the original two. Somehow or other the attractions of railways had now lost their appeal, and passing regularly through this station is now only a necessary part of rail travel.

6TH JUNE 1957 • It was not just frequent passenger services which made 'train spotting' at Coventry so attractive, much heavy freight also trundled over the station metals as depicted here whilst a tank engine enters the main line through the station with a set of coaches from the Leamington direction. Resting at the 'Up' platform are two Stanier '8F' Class 2-8-0s with a long train of coal empties. The leading engine is No. 48723, believed to be based at Nuneaton MPD (2B). PHOTO: MIKE MITCHELL (MM281) © THE TRANSPORT TREASURY

6TH JUNE 1957 • A rare visiting special – the '1907-1957 TT Motor Cycling's Excursion Golden Jubilee' – rests in Coventry station with Stanier Black 5 4-6-0 No. 44863 at the head of the train. PHOTO: MIKE MITCHELL (MM282) © THE TRANSPORT TREASURY

24TH OCTOBER 1959 • Coventry station west 'Up' platform had a small goods yard; on this occasion hosting a former LMS parcels van. Simmering away at the 'Down' platform is 'Black 5' 4-6-0 No. 45083, having just arrived with a crowded passenger train from the south. The period lower quadrant signal really sets the scene. PHOTO: MIKE MITCHELL (MM282) © THE TRANSPORT TREASURY

21ST MAY 1957 • Deputising for a possible 'failed' 'Jubilee', double-headed Stanier Black Fives with No. 45257 leading, at speed near Willenhall on the eastern suburbs of Coventry with the 'Down' THE MIDLANDER express. PHOTO: MIKE MITCHELL (MM251) © THE TRANSPORT TREASURY

6TH JUNE 1957 • Stanier 4-6-0 power in abundance hauling the 'Down' THE MIDLANDER express into Coventry station; headed by Rugby (2A) based Class 5 No. 44909 with double-headed assistance by an unidentified 'Jubilee', which may well have been the principal means of motive power. PHOTO: MIKE MITCHELL (MM283) © THE TRANSPORT TREASURY

24TH OCTOBER 1959 • Whilst this train cannot be identified as the 'Up' THE MIDLANDER express, such a train was invariably hauled by a Stanier 'Jubilee' Class of locomotive. On this occasion 45734 METEOR stands at Coventry station before living up to its name, and heading off at speed to its destination. PHOTO: DICK RILEY (RCR14403) © THE TRANSPORT TREASURY

24TH OCTOBER 1959 • Nuneaton (2B) Stanier Class 8F 2-8-0 No. 48312 proceeds through the main line of Coventry station before taking the Kenilworth curve with what looks like a heavy coal train probably from the nearby coalfields. Meanwhile, the 'Up' platform loop is still clear for the train having left there. PHOTO: DICK RILEY (RCR14406) © THE TRANSPORT TREASURY

Hughes Class 6P5F 'Crab' mogul No. 42923 pulls into Coventry station's 'Down' platform loop with a train from the Leamington direction.
PHOTO: MILEPOST 92½ (MC30123) © THE TRANSPORT TREASURY

24TH OCTOBER 1959 • Coventry No. 2 Signal Box, provided by Saxby & Farmer in 1874, at the western end of the main station, with part of the parcels depot located where the coach is stabled. For many years the rear of the London & Birmingham station building was screened from view by the roof of the parcels dock which stood in front of it. PHOTO: DICK RILEY (RCR14410) © THE TRANSPORT TREASURY

1st September 1962 • Coventry had a substantial goods yard at the west end of the station. In this view Willesden (IA) based Class 8F 2-8-0 No. 48122, with guards van attached, awaits its next duty. PHOTO: TONY COUSINS (TC247) © THE TRANSPORT TREASURY

1st September 1962 • Standard Pacific 'Britannia' Class No. 70053 MORAY FIRTH joins the main line with its through train to Birmingham and passes Coventry goods yard, having just left the station loop. PHOTO: TONY COUSINS (TC246) © THE TRANSPORT TREASURY

26TH DECEMBER 1960 • Ivatt 'Mucky Pig' class 4MT 2-6-0 No. 43002, based at Nuneaton (2B) shed, makes a lot of smoke with a local train heading towards Birmingham. Its home shed route verges off to the left beyond the 'Home' and 'Distant' signal gantry.
PHOTO: TONY COUSINS (TC071) © THE TRANSPORT TREASURY

24TH SEPTEMBER 1959 • Coventry station complex looking west towards Birmingham, with a Class 101 DMU waiting in the 'Up' platform, whilst its signal clears for that train to proceed. Beyond the engine shed turntable to the left are the tracks to Leamington, whilst Coventry No. 1 signal box stands guard over all the lines. Below shows the adjacent scene from Quinton Road bridge, overlooking the engine shed. What a modelling inspiration. PHOTO: DICK RILEY (RCR14404/05) © THE TRANSPORT TREASURY

3RD SEPTEMBER 1962 •
Ivatt Class 4MT No. 43002, a design introduced in 1947 and based at Nuneaton (2B) shed, rests for the time being in Coventry MPD (2D). The tracks linking Coventry and Leamington lie to the left of the locomotive.
PHOTO:
TONY COUSINS (TC249)
© THE TRANSPORT TREASURY

1949 • Under its new masters and still wearing its early BR livery, Fowler Class 3 2-6-2T No. 40002 of Warwick (2E until 1950, being recoded 2C) is pictured in light steam at Coventry. It was based at Warwick until April 1953 when transferred to Willesden (1A).
PHOTO: NEVILLE STEAD COLLECTION (NS203549)
© THE TRANSPORT TREASURY

5TH FEBRUARY 1953 •
Another Fowler Class 3 2-6-2T, this time No. 40016, had recently been transferred to Willesden (1A) from Warwick (2C), so may have been laying over at Coventry MPD on its journey south.
PHOTO:
ERIC SAWFORD (ES853)
© THE TRANSPORT TREASURY

5TH FEBRUARY 1953 • All fired up and somewhere to go, after final attention from the driver. Ivatt Class 2 2-6-2T No. 41234 is prepared for the road ahead, perhaps returning to its home shed at Nuneaton (2B). PHOTO: ERIC SAWFORD (ES856) © THE TRANSPORT TREASURY

5TH FEBRUARY 1953 • Steam in abundance at Coventry shed, with Rugby (2A) allocated Fowler Compound 4-4-0 No. 41162 prominent. Put into service by the LMS in September 1925, withdrawn by BR in June 1960. PHOTO: ERIC SAWFORD (ES852) © THE TRANSPORT TREASURY

5TH FEBRUARY 1953 • Bowen-Cooke Class 7F 0-8-0 No. 49425 belches out plenty of smoke whilst it is being loaded at Coventry coaling plant. Once allocated to Rugby (2A) depot, it would find itself moved up the road to Nuneaton (2B) for similar duties.
PHOTO: ERIC SAWFORD (ES855) © THE TRANSPORT TREASURY

MIDLAND TIMES • ISSUE 3

# WESTHOUSES YARD, DERBYSHIRE
## Fowler 4F nos. 43850 and 44130 – 18th September 1964

## MIDLAND PULLMAN AT AMBERGATE, DERBYSHIRE
### 29th September 1964

## MIDLAND TO THE CORE: JAMES WILLIAM WATKINS
### by Graham Rankin

Ashchurch Station, on the Birmingham and Gloucester Railway, was opened in 1840. It was once a minor railway centre because it was the spring-off point for two branch lines either side of the main line. One of my distant relations was the station master there.

William Watkins (1858-1922) was Ashchurch Station Master between 1900 and 1922. Previously, he had been Station Master at Bromsgrove in Worcestershire. He had two sons, both of whom joined the Midland Railway as clerks at Ashchurch. Frederick Charles Watkins (born 1886) worked as a clerk at various stations in Gloucestershire until at least 1927. His brother had a more illustrious career.

James William Watkins was appointed to the Midland Railway in 1905 as a clerk at Ashchurch Railway Station. By 1914, he was a booking clerk in the Traffic Department (Coaching) at Great Malvern in Worcestershire.

On the outbreak of World War One, James enlisted as a Private in The Gloucestershire Regiment in August 1914. He did not take advantage of his protected status as a railwayman to avoid military service. James had a distinguished war record and progressed through the ranks. By the end of the war, he was a Lieutenant Colonel who had been awarded the Distinguished Service Order and the Military Cross for bravery in action. In August 1918, he married Ethel Mary Price (born 1894) at Tewkesbury Wesleyan Church in Gloucestershire.

By 1922, he was living in Derby where he was appointed to the newly-created London, Midland and Scottish Railway (LMS) in some sort of managerial capacity, given his wartime experience. By 1938, he was Assistant Divisional Superintendent of Operations.

In 1938, James attended the funeral of Sir Henry Fowler (1870-1938) who had been Chief Mechanical Engineer of the Midland Railway (1909-23) and the LMS (1925-31).

A superb panoramic view of Ashchurch station on the Midland Railway's Bristol to Birmingham main line looking north c.1960. An unidentified Standard Class 5 makes a spirited departure. The line to the left is to Tewkesbury and that to the right towards Evesham.
Photo: Milepost 92½ (MPM996-00047) © The Transport Treasury

In 1942, James was appointed to the position of Divisional Superintendent of Operations at Crewe. Then in 1948, he became Operating Superintendent of the newly-formed London Midland Region of British Railways (BR).

On 15th September 1948, he attended a lunch at Euston Station Hotel with honoured guest, Field Marshall, The Viscount Montgomery KG, GCB, DSO. The occasion was the naming of LMS Patriot No. 45506 as *The Royal Pioneer Corps*. He was one of only three railway representatives there.

In 1951, James was promoted to Chief Regional Officer of the London Midland Region, BR.

In the King's Birthday Honours List, he was appointed Commander of the Royal Victoria Order. This is awarded usually for distinguished personal service to the monarch. However, I have not been able to find out what that entailed.

On October 1953, the Railway Executive was abolished and James role became General Manager, London Midland Region, British Railways. In March 1953, James was the official guest at the Institution of Locomotive Engineers' Dinner at the Dorchester Hotel in London.

During James' tenure as General Manager, the BR Modernisation Plan was adopted. He was instrumental in the introduction of lightweight Diesel Multiple Units in West Cumberland. James retained the General Manager role until his retirement in 1956. Thereafter, he became a full-time member of the British Transport Commission: he was one of only six individuals reappointed in a full-time role. He also became Chair of the Railway Study Association and Vice-President of the Institute of Transport.

The remarkable working life of James William Watkins spanned the years from pre-grouping to nationalisation. It was interrupted only by distinguished military service. His career took him from appointment as a lowly clerk to executive positions in BR and other organisations. It is clear from the preceding narrative that there are gaps in my knowledge of his career. Perhaps other subscribers to this magazine will be able to fill some of them? If you can help with any further information please email: *midlandtimes1884@gmail.com*

### Website sources:

*British History Online – Parishes: Ashchurch*
*Disused Stations: Ashchurch*
*Gloucester Railway Memories: Ashchurch – A Country Junction*
*Midland Railway Society: Ashchurch Station Masters*

Fowler Class 3P 2-6-2T No. 40040 awaits departure time at Ashchurch with its single carriage service for Tewkesbury.
PHOTO: NEVILLE STEAD COLLECTION (NS203702) © THE TRANSPORT TREASURY

One of only three railway personnel present, Watkins attended the naming ceremony of Patriot Class 5XP 4-6-0 No. 45506 THE ROYAL PIONEER CORPS which looks pristine in this September 1948 works photo in its newly-applied BR black lined livery.
PHOTO: COURTESY, THE LMS-PATRIOT PROJECT

Watkins was instrumental in the introduction of DMUs in Cumbria. Seen here is Derby Lightweight DMU with Driving Trailer Open Lavatory Composite No. M79602, awaiting departure from Carlisle Citadel with a service for Whitehaven on 23rd April 1957.
PHOTO: NICK NICOLSON (LN729) © THE TRANSPORT TREASURY

# GARSDALE

Garsdale station on the Settle to Carlisle line is located at a bleak spot, where winds funnelled by the fells can meet from two different directions at the same time. Around the station is a scattered community with a line of former railway staff cottages, a number of isolated farms, a Mount Zion Chapel, and the Moorcock Inn.

Originally called Hawes Junction, there was a six mile line to the Wensleydale town of Hawes. Here the line became the North Eastern line to Northallerton. This part of the line was included in Contract No. 2 (Dent Head to Kirkby Stephen) that was awarded to Messrs. Benton and Woodiwiss, who were also contracted to build the branch line to Hawes. For this they estimated that around 1,400 men would be needed, in this they were successful but due to the harsh working conditions there was a high turnover of workers.

The waiting room of the northbound platform was used for church services before the second world war, the ladies waiting room contained a library of 150 books, while the stone base of the water tower was used as the local dance hall, cinema, concert hall and social centre. It could house around 50 to 60 people, it was also home to domino and whist drives, suppers and was used on many occasions for wedding and birthday parties. The tank above had a capacity of 90,000 gallons of water which was not only used to replenish the locomotives but, via an open spout, fed the station and cottages.

Many of the original Midland Railway fittings have been removed, notably the turntable which was located in a particularly exposed position; it was not unknown for engines to be spun around in ferocious winds. Due to this it was provided with an airtight fence of upright sleepers. The turntable has been restored and installed at the Keighley and Worth Valley Railway.

Garsdale station is different from other stations on the line as the station buildings do not use the standard station building design as used at all other stations. In 1872 the Midland Railway envisaged a small type station but in fact the requirements at this station resulted in three buildings along the same design as the waiting rooms at other stations. It was also the only junction station on the Settle & Carlisle.

Occupying the downside platform (towards Carlisle) is Garsdale signal box, brought into use in 1910. This replaced the original north and south boxes. It was larger than most of the other signal boxes on the line, it contained a 40 lever tappet frame and was in use until 1983.

Opened on 1st August 1876 as Hawes Junction, adjoining the station are sixteen Railway Cottages built for its employees by the Midland Railway around 1876, the year the Settle-Carlisle Line opened. A further six cottages were added near to the Moorcock Inn soon afterwards. Original plans had outlined a complex on a larger scale which was intended to be an exchange station with sidings for goods traffic and an engine shed large

*Garsdale viewed from the south, with the Hawes line branching off to the right.*
PHOTO: RON SMITH © THE TRANSPORT TREASURY

enough to accommodate 24 engines with no less than 30 cottages built to house staff.

The first service over the Settle to Carlisle line occurred on 1st May 1876 hauled by Kirtley loco No. 806. Compound 4-4-0 locos became the mainstay of passenger services, replacing smaller Midland Railway engines that required double-heading over the heavily inclined line, these were then superseded in subsequent years by Black Fives, Jubilees and Rebuilt Royal Scots.

To the north of the station stands Dandry Mire Viaduct (also known as Moorcock Viaduct or Garsdale Viaduct), which crosses Dandry Mire, and the vantage point just north of Garsdale, overlooking Dandry Mire, is one of the most popular photographic positions on the railway. Originally planned as an embankment, problems were encountered. After several months of earth being tipped, this continued to sink into the moss leading, in 1872, to the draining of the embankment and a trench being cut into the peat to find solid bedrock, some 50 feet down, for the foundation of the viaduct. The result was a 227 yard long curved stone structure which contained 12 arches varying in width from 44ft. 3in. to 45ft.

Located to the south of the station and just to the north of Rise Hill tunnel were Garsdale water troughs. The troughs were installed in 1907 at a cost of £4,396. They were 1,670 feet in length and held between 5,000 and 6,000 gallons of water, of which up to a third would be taken in by a tender in a matter of seconds. They were fed from a 43,000 gallon tank that received its supply by pipe from a reservoir in the hillside above; work on the dam and reservoir took around 12 months to complete and required a workforce of 50 to 60 men. Additional liquids were fed into the tank to keep the water clean and free-running, to soften the water and to prevent the interior of the boiler from rusting. In winter, the tank was kept heated by a boiler. Steam passed through copper pipes. A man on night-duty stoked up the fire using, for some of the time, coal that had been washed from the tenders of passing trains.

Between 1st August 1878 and 16th March 1959, it was the junction between the Hawes branch and the main Settle to Carlisle railway. When the station first opened, it was known as Hawes Junction. In January 1900, the name changed to Hawes Junction and Garsdale; in September 1932 the name was shortened to Garsdale, thereby losing the reference to the junction.

### References:
*Stations and Structures of the Settle to Carlisle Railway:*
V. R. Anderson and G. K. Fox.

*Garsdale:* W. R. Mitchell

**9TH JULY 1961** • Garsdale looking north towards Carlisle, showing the palisade around the turntable and the Hawes branch trailing off to the right.
Photo: Alan Robey © The Transport Treasury

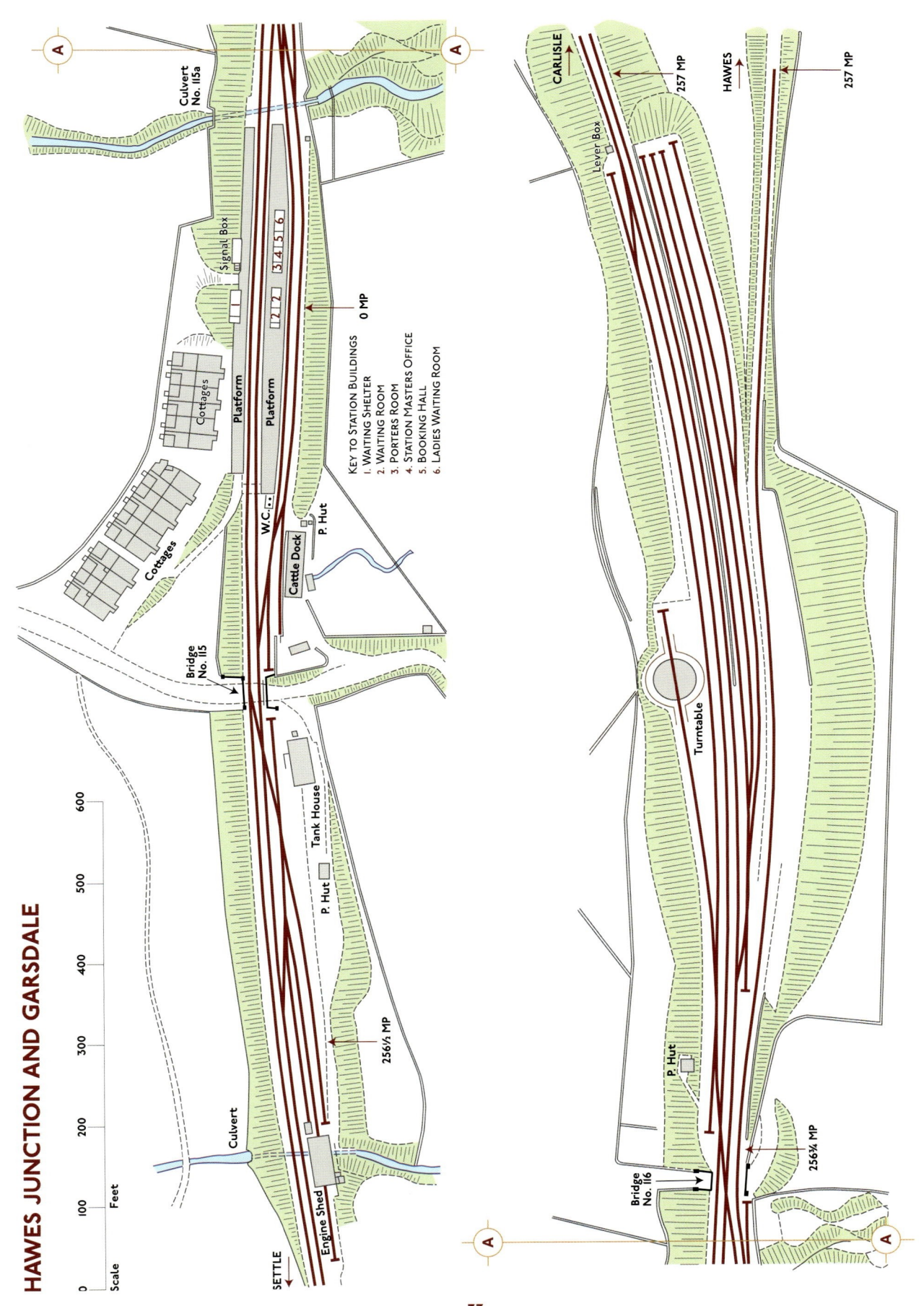

# HAWES JUNCTION AND GARSDALE

**Key to Station Buildings**
1. Waiting Shelter
2. Waiting Room
3. Porters Room
4. Station Masters Office
5. Booking Hall
6. Ladies Waiting Room

The view south from Garsdale station with the water tower prominent on the left.
PHOTO: RON SMITH © THE TRANSPORT TREASURY

Fairburn Class 4MT 2-6-4T No. 42132 pulls into Garsdale on a local service in 1960. The photo gives us a good view of the 1910-built signal box that replaced boxes to the north and south of the station.
PHOTO: NEVILLE STEAD COLLECTION (NS209359) © THE TRANSPORT TREASURY

The Thames-Clyde Express heads south through Garsdale with Rebuilt 'Royal Scot' Class 7P 4-6-0 No. 46145 The Duke of Wellington's Regiment (West Riding) at the head of the train.
PHOTO: NEVILLE STEAD COLLECTION (NS209361) © THE TRANSPORT TREASURY

The palisaded turntable at Garsdale shown to good effect as Ivatt Class 2MT 2-6-2T No. 41206 is being turned.
PHOTO: NEVILLE STEAD COLLECTION (NS209358) © THE TRANSPORT TREASURY

A track level view of Jubilee Class 6P 4-6-0 No. 45565 VICTORIA calling at Garsdale in 1954 with a local service.
PHOTO: NEVILLE STEAD COLLECTION (NS209360) © THE TRANSPORT TREASURY

With Dandry Mire viaduct in the background, Stanier 8F 2-8-0 No. 48542 approaches Garsdale station light engine on a bleak winter's day. Note that both the turntable and Hawes branch line have been removed.
PHOTO: NORRIS FORREST (NF231-05) © THE TRANSPORT TREASURY

A 1950s view of Garsdale troughs showing a Carlisle-bound freight with a Hughes/Fowler Class 5MT 2-6-0 'Crab' in charge.
PHOTO: MILEPOST 92½ (MP90263) © THE TRANSPORT TREASURY

Another 1950s view of Garsdale troughs, this time looking north.
The 43,000 gallon tank that fed the troughs can be clearly seen on the right.
PHOTO: MILEPOST 92½ (MP90264) © THE TRANSPORT TREASURY

# RECREATING A COLOURFUL PAST
## The images of David P. Williams

The cover on the first three issues of Midland Times featured the work of David P. Williams, whose knowledge of railways enable him to produce superb coloured monochrome images. These images allow us to see what a colourful and vibrant railway we once had.

In 1953 my parents bought their first television. The stimulus for them and many others at the time was the forthcoming Coronation which was to be televised across the nation.

It is hard to appreciate in the 21st Century just how primitive the viewing experience was in those days, as the digital age was still a long way into the future. Fuzzy images accompanied by varying amounts of 'snow' on the screen could sometimes be improved by slightly adjusting the position of the aerial – but this was also done at risk of making things worse. The picture was monochrome of course as we lived in a world of monochrome photography in those days, and were well used to seeing images devoid of colour in books, magazines and at the cinema. Ten years later at the start of a degree course, I was whisked from the backwaters of Teesside to London, where colour television was being experimentally broadcast.

Within days of my arrival, walking along Chiswick High Street, my attention was drawn to a group of bystanders gazing in a shop window at a small television broadcasting in full colour. It was an experience which remains vivid in my memory. We take it all for granted now, but colour images were not always so instantly and freely accessible.

Although colour television did not come into its own until the 1960s, colour photography had been around since the mid-1930s, and grew rapidly after the war. In the same way that the early colour television experience bore a pale comparison to that of today, the same is true of early colour photography which was a very expensive activity and limited in its scope. Development of colour photography with improving film speeds, greater availability and reduced cost gathered pace during the 1950s but it remained something of an elitist branch of the art until the advent of digital photography.

Railway photography has a long and distinguished past which has left a rich legacy of images to enjoy, but the fact remains that most of these photographs were taken in the days of monochrome film. This is something of a tragedy as history tells us just how colourful the railway scene used to be. Before the Grouping a host of different companies vying for the traveller's custom used distinctive liveries as part of their armoury in attracting patronage. Even in the days of the 'Big Four' colour was a vital part of the commercial strategy, so clearly demonstrated by the inspirational choice of garter blue by the LNER for its A4 pacifies.

We are now able to recreate the past by digitally adding colour to monochrome pictures. News coverage was recently given to cine film of World War 1 originally shot in monochrome, which has been dramatically brought to life in this way. The technique is not without its detractors who point to the subjectivity in choice of the applied colours. There is no answer to this as these selections must be made somehow, and there are literally millions of colours, shades and intensities available. The die-hards point to the objectivity of colour transparencies whilst conveniently forgetting that analogue colour film was notorious for variability in its results depending on the make of film and type of colour processing. We have a simple choice to reject or embrace such coloured images, but to reject them is to deny the possibility of enjoying colourful scenes that existed in the past. Whoever attempts the transformation bears a heavy responsibility to get things right and do the necessary homework. The health warning that comes with a coloured monochrome picture is no different from that which came with colour transparencies – there are no guarantees with this product.

As with conventional photography it is a skill which must be acquired the hard way, in my case with years of painstaking trial and error. There are no simple textbooks. It helps to have been an active railway photographer, as the colour differences between rails and wooden sleepers, both shades of brown, become ingrained in the memory. The object of the exercise is always the same, to recreate the scene observed in the viewfinder by the original photographer. Sometimes there are elements which simply cannot be known. A woman stands on the platform wearing a dress which appears as dark grey on the original. It could have been dark grey but could also have been various dark shades of red, blue, green, etc. Some colours can be ruled out – yellow for example. The original monochrome picture sets its own limits for what is possible. The die-hard will crow that the colour transparency despite its deficiencies will give a better idea of the woman's dress colour. True, but there is another important advantage of coloured photographs to be mentioned in passing, which concerns picture resolution. The vast majority of colour transparencies in the Railway Archive were taken on 35mm film or smaller, sometimes using a camera of lesser quality. When such pictures are enlarged the limited resolution often becomes starkly revealed, but it need not be this way for the coloured monochrome photograph. The amount of time and effort expended on one work makes it necessary to choose the original carefully. A really sharp, high quality large format negative can result in a colour picture which retains good resolution to a huge magnification, sometimes with truly breathtaking results.

The proof of any pudding is in the eating. We all have a concept of correctness and are free to make our own judgement on whether any coloured photograph is an accurate representation of reality. The examples on the front cover, centre spread and ex-works No. 1119 seen at Crewe in May 1937 (below) provide scope for debate.

# 6202 at Camden Shed

An early 1936 portrait of the unique, experimental 'Turbomotive' shows No. 6202 in original condition at Camden shed in the company of 'Royal Scot' No. 6133 VULCAN. The following year, a higher superheat domed boiler was fitted to No. 6202 and, in 1939, smoke deflectors were added. Apart from being driven by Metropolitan Vickers turbines, the locomotive was otherwise conventional with a coal burning firebox, firetube boiler and superheating elements. Rebuilt in 1952 as a non-turbine driven locomotive and named PRINCESS ANNE, the 4-6-2 ran for only two months before it was wrecked in the Harrow & Wealdstone disaster on 8th October 1952, and subsequently scrapped.
© DAVID P. WILLIAMS COLOUR ARCHIVE

# LMS GENERAL INSTRUCTIONS AND DIAGRAMS

The following selected illustrations are from an official LMS document produced in 1941 for guidance of staff in loading and securing long and projecting traffic, to ensure safe conveyance to their destination on the LMS system or exchanged with other railway companies.

Thank you to Jeremy Clements for donating the booklet for publication in Midland Times.

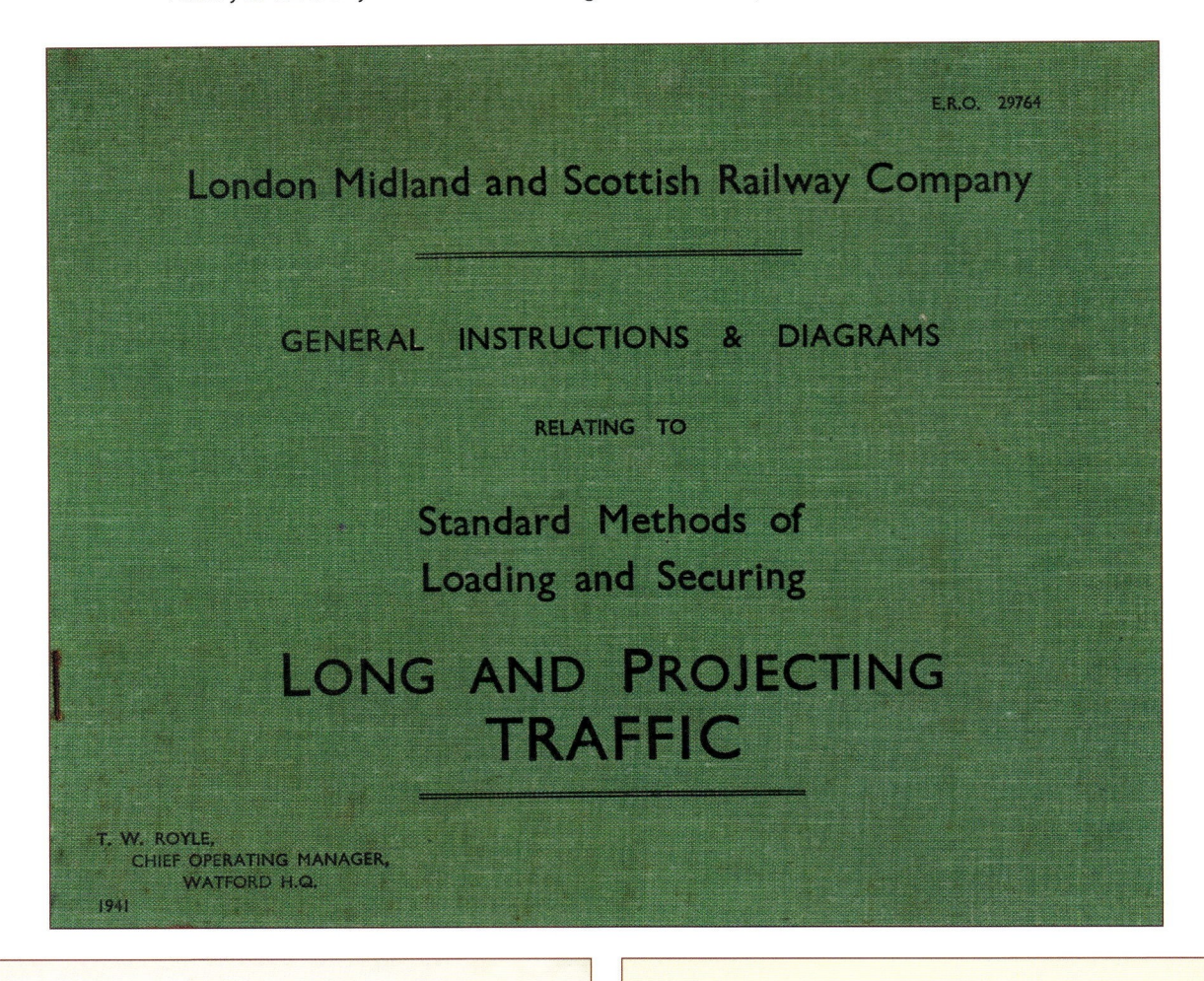

---

**E.R.O. 29764**

**London Midland and Scottish Railway Company**

GENERAL INSTRUCTIONS & DIAGRAMS

RELATING TO

Standard Methods of Loading and Securing

## LONG AND PROJECTING TRAFFIC

T. W. ROYLE,
CHIEF OPERATING MANAGER,
WATFORD H.Q.
1941

---

**LONDON MIDLAND & SCOTTISH RAILWAY COMPANY**

General Instructions and Diagrams
Relating to
Standard Methods of
Loading and Securing
Long and Projecting Traffic

1941

Whilst these instructions include loading arrangements for long and projecting traffic conveyed on sets of Single Bolster Wagons, preference should be given to making self-contained loads on Double Bolster, Bogie Rail and Bogie Timber Wagons or other suitable stock whenever practicable.

All concerned are reminded that loads exceeding 60 feet in length, or more than 30 feet from centre of carrying wagon of self-contained loads or from mid-way between securing points in the case of sets of Single Bolster Wagons, to either end, are subject to special arrangements for transit under the standard regulations governing the conveyance of out of gauge and otherwise exceptional loads contained in the General Appendix to the Working Time-Tables, and in the Commercial and Operating Departments Joint Circular dated 12th September, 1936.

These instructions, with the accompanying diagrams, are for the guidance of staff in loading and securing long and projecting traffic, to ensure safe conveyance, whether arising on the London Midland and Scottish Railway system or exchanged from other Railway Companies, and must be observed.

Goods Agents or other persons in charge of Depots must see that the Company's loading staff and others concerned understand the instructions.

(1) Before loading see that :—
   (a) Chains and Stanchions are in a satisfactory condition.
   (b) Swivel bolsters of single bolster wagons work freely.
   (c) Tops of bolsters in sets of single bolster wagons are equal in height from rail level.

(2) Carrying Wagons must not be loaded in excess of their registered capacity, or exceed that where shewn under diagrams.

---

(3) Where bolsters of Single Bolster Wagons are not provided with iron plates for conveyance of iron and steel-work, the bolsters on which the load is free to move must be well greased.

(4) All loads on sets of Single Bolster Wagons must overlap the bolsters at ends by at least 2 feet 6 inches.

(5) Sets of Single Bolster Wagons should be so coupled as to limit the maximum possible movement between any two vehicles to 1 foot 4 inches.

(6) Bolsters removed from their normal position in under-running wagons should be disconnected and laid in a safe place in the wagon clear of the load; care must be taken to see that the pivot pins accompany the bolsters. Stanchions and Bolsters to be replaced in their normal positions when vehicles are running light.

(7) Loads must have at least 4 inches clearance above floors, sides and ends of Single Bolster Wagons, including match, guard or runner wagons.

(8) Wagons used as match, guard or runner wagons which are fitted with bolsters must have the bolsters removed unless there is a clearance of at least 4 feet at either end. Bolsters removed to comply with this requirement should be placed at the end of the wagon away from the load.

(9) In loading timber of unequal lengths, longer pieces should, as far as possible, be placed at the bottom and in the centre of the load, shorter pieces on top and all firmly bound together with chains and screw shackles.

(10) Whenever practicable, straight trees should be loaded at the bottom and those of irregular shape on top of load.

(11) Trees and poles which are thicker at one end than the other loaded on sets of Single Bolster Wagons must be tightly secured with chains at the butt ends, but at the other bearing and securing point the chains must be applied so as to permit of slight backward and forward movement.

(12) Chains are preferable for binding tapering ends of round timber, but where chains are unobtainable, ropes may be used, provided a satisfactory binding can be made.
   Note : This binding must be free of the wagon.

Where doubt exists as to the safety of loads, the District Goods Manager or District Traffic Superintendent concerned (Operating Manager, Northern Division) must be consulted and his instructions obtained.

**DIAGRAM NO. 6A**

## METHODS OF CARRYING TELEGRAPH POLES.
(Tip to Butt Loading)

### 3 WAGON SET

MAY BE LOADED TO FULL WIDTH OF BOLSTER. STANCHIONS TO BE REMOVED FROM CENTRE WAGON AND LOAD CHAINED FREE OF WAGON.

**TOTAL WEIGHT OF LOAD NOT TO EXCEED CAPACITY OF TWO WAGONS.**

A.—LOAD CHAINED TO BOLSTERS, STANCHIONS IN POSITION, SHACKLES **OUTSIDE** STANCHIONS WHEN, OWING TO LOAD EXTENDING TO FULL WIDTH OF BOLSTERS, THEY (THE SHACKLES) CANNOT CONVENIENTLY BE PLACED **INSIDE** THE STANCHIONS.

B.—LOAD CHAINED FREE OF WAGON

C.—BOLSTER TO BE RETAINED.

### SELF CONTAINED LOAD

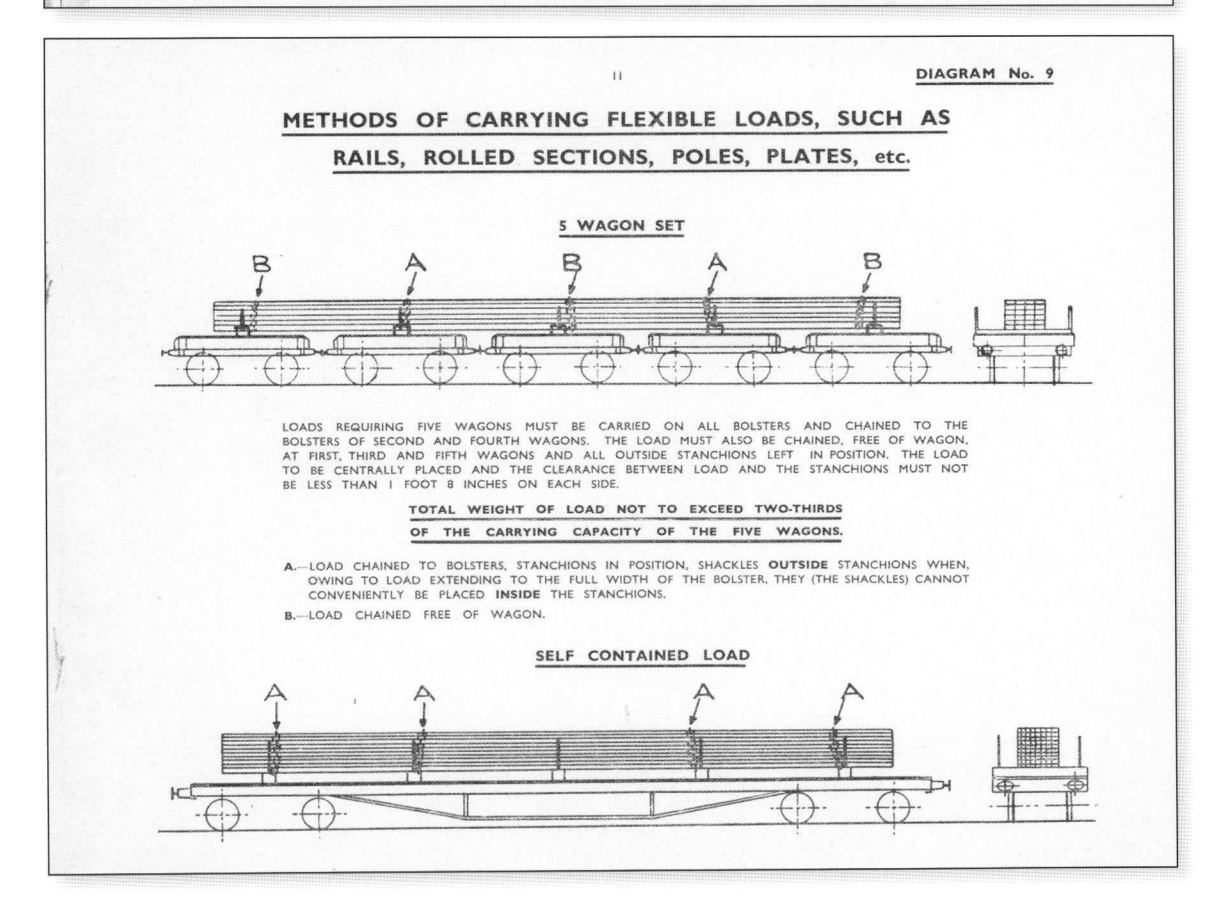

---

**DIAGRAM No. 9**

## METHODS OF CARRYING FLEXIBLE LOADS, SUCH AS RAILS, ROLLED SECTIONS, POLES, PLATES, etc.

### 5 WAGON SET

LOADS REQUIRING FIVE WAGONS MUST BE CARRIED ON ALL BOLSTERS AND CHAINED TO THE BOLSTERS OF SECOND AND FOURTH WAGONS. THE LOAD MUST ALSO BE CHAINED, FREE OF WAGON, AT FIRST, THIRD AND FIFTH WAGONS AND ALL OUTSIDE STANCHIONS LEFT IN POSITION. THE LOAD TO BE CENTRALLY PLACED AND THE CLEARANCE BETWEEN LOAD AND THE STANCHIONS MUST NOT BE LESS THAN 1 FOOT 8 INCHES ON EACH SIDE.

**TOTAL WEIGHT OF LOAD NOT TO EXCEED TWO-THIRDS OF THE CARRYING CAPACITY OF THE FIVE WAGONS.**

A.—LOAD CHAINED TO BOLSTERS, STANCHIONS IN POSITION, SHACKLES **OUTSIDE** STANCHIONS WHEN, OWING TO LOAD EXTENDING TO THE FULL WIDTH OF THE BOLSTER, THEY (THE SHACKLES) CANNOT CONVENIENTLY BE PLACED **INSIDE** THE STANCHIONS.

B.—LOAD CHAINED FREE OF WAGON.

### SELF CONTAINED LOAD

**DIAGRAM NO. 13A**

## MAXIMUM LOADS AND OVERHANGS OF TIMBER, PIPES, ANGLES, BARS, etc., NOT LOADED TO FULL WIDTH OF WAGON, WHICH MAY BE CONVEYED IN ORDINARY 8, 10 AND 13 TON WAGONS

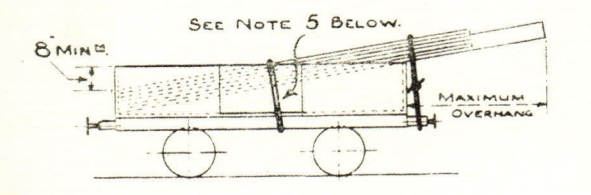

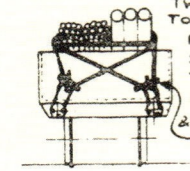

See Note 5 Below. 8 Min. MAXIMUM OVERHANG

TWO ROPES TO BE USED TO SECURE THE OVERHANG. ROPES TO BE FIRST SECURED TO LOAD, THEN DOUBLE ROPED ROUND BUFFER CASTING & FINALLY SECURED AS SHOWN

### TABLES FOR LOADS OVERHANGING ONE END

| Maximum Overhangs with Equal Lengths (Solid Load) | 8 ton wagon Steel Frame | 8 ton wagon Wood Frame | 10 ton wagon Steel Frame | 10 ton wagon Wood Frame | 13 ton wagon Steel Frame | 13 ton wagon Wood Frame |
|---|---|---|---|---|---|---|
| 1 ft. | 5 | 5 | 6 | 6 | 6 | 6 |
| 2 ft. | 4 | 4 | 5 | 5 | 6 | 6 |
| 3 ft. | 4 | 3 | 5 | 4 | 6 | 5 |
| 4 ft. | 3 | 2½ | 5 | 3 | 6 | 4 |
| 5 ft. | 2 | 2 | 4 | 2½ | 5 | 3 |
| 6 ft. | 2 | 1½ | 3 | 2 | 4 | 2½ |

| Max. overhangs with unequal Lengths (not Exceeding 50% of Load to be solid.) | 8 ton wagon Steel Frame | 8 ton wagon Wood Frame | 10 ton wagon Steel Frame | 10 ton wagon Wood Frame | 13 ton wagon Steel Frame | 13 ton wagon Wood Frame |
|---|---|---|---|---|---|---|
| 6 ft. and under | 4 | 3 | 5 | 4 | 6 | 5 |
| 7 ft. | 3 | 2½ | 5 | 3½ | 6 | 4½ |
| 8 ft. | 3 | 2½ | 4 | 3 | 6 | 4 |

NOTES:—
1. Length of load must in no case exceed length of wagon body by more than 8 feet. (See note 3 for end-door wagons).
2. Any load exceeding 25 feet 6 inches in length must be loaded flat on other suitable stock other than ordinary wagons.
3. End-door wagons may be used providing the end doors are properly secured. The overhanging portion must rest on the door hinge bar and not exceed 2 feet nor the maximum load 2 tons.
4. Overhang to be at trailing end where possible.
5. Ropes must be used to secure the overhanging end of the load as indicated and the additional rope shown at the centre of the wagon is to be used when the load is above the rave of the wagon at this point.
6. Loads not to overhang less than 9 inches.
7. Loads must be laid longitudinally and not diagonally across the wagon.

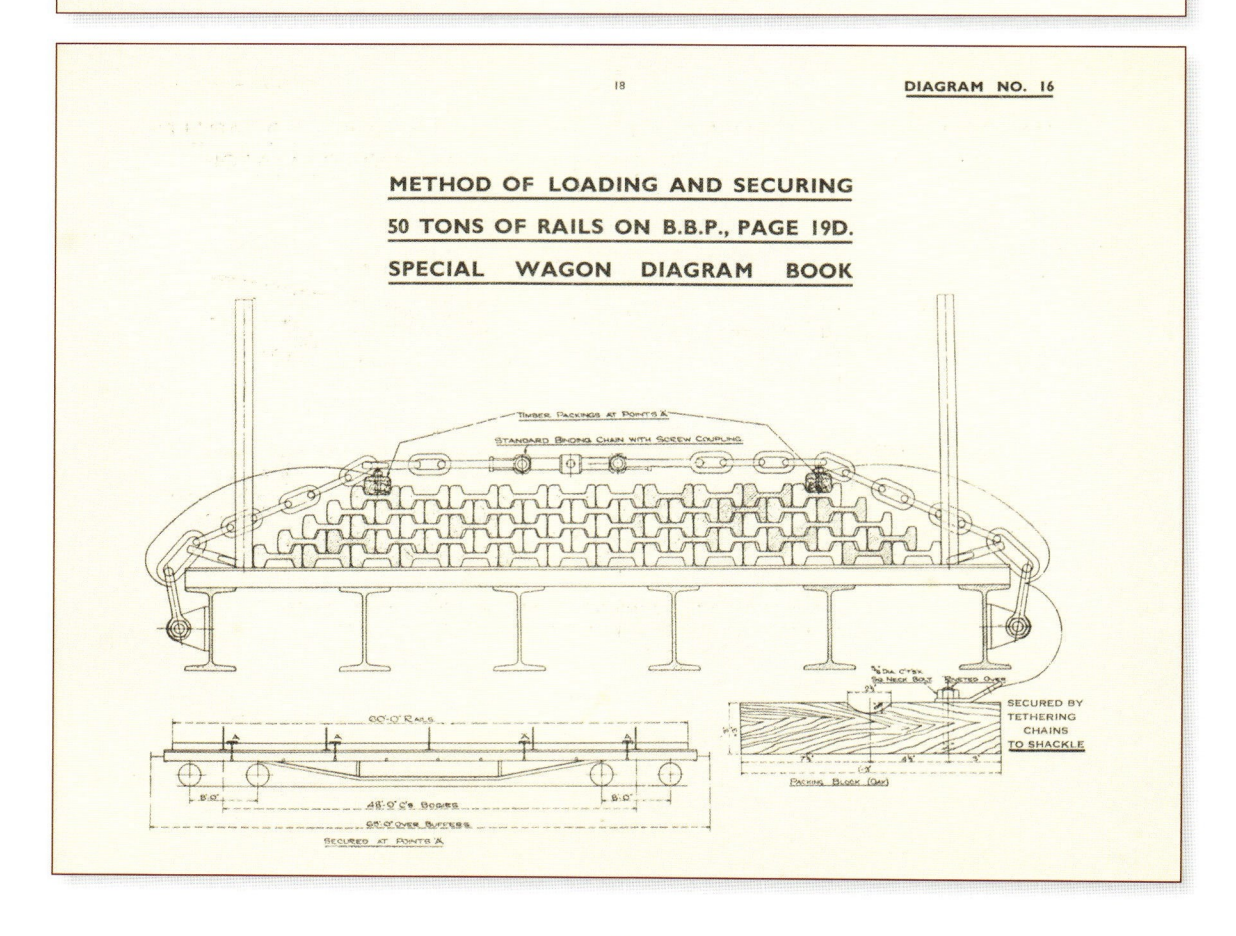

**DIAGRAM NO. 16**

## METHOD OF LOADING AND SECURING 50 TONS OF RAILS ON B.B.P., PAGE 19D. SPECIAL WAGON DIAGRAM BOOK

19          DIAGRAM NO. 17

## METHODS OF LOADING & SECURING 90-ft. RAILS

"A."—Bolsters removed.      "B."—Bolsters plated and greased.

"C."—Securing bolsters with timber packing ½″ in excess of thickness of plates on bolsters "B" and load to be double chained.

"D."—Load chained free of wagon.

"E."—Stanchions in inner sockets (where provided) of securing bolsters "C."

"F."—Load chained to bolsters.

*Note :—L. & N.E. Co.'s 40 Ton (old type) Quintuple bolster wagons with 5 fixed bolsters (bolsters not removeable) may be used.

SEE NEXT PAGE

---

20          DIAGRAM NO. 17

**Loading and Securing Rails 90 feet in length.**

The following instructions must be observed in connection with the loading and securing of rails 90 feet in length.

**Types of Wagons.**

When rails 90 feet in length cannot be loaded to make a self-contained unit (No. 1 on diagram) they must be loaded with two similar bogie bolster wagons (Nos. 2, 3 or 4 on diagram). The Wagons must not be less than 45 feet nor exceed 52 feet in length over headstocks, and must have fixed bolsters of equal height from rail level.

**Weight of Loads.**

The maximum weight must not exceed half the combined carrying capacity of the two wagons, and in no case exceed 30 tons.

The weight must be evenly distributed over the two wagons, and as far as practicable, the load must be equal ended and central.

**Securing of Loads.**

The securing must be at the centre bolster of each wagon, where centre bolsters are provided ("C" diagrams 2 and 3) or, in the case of four-bolster type wagons, at the second bolster from the ends of the load ("C" diagram 4).

Suitable steel plates (skids), well greased, must be fixed on bolsters other than securing bolsters ("B" diagrams 2, 3 and 4).

Packing on securing bolsters ("C" diagrams 2, 3 and 4) must be of timber approximately ½″ thicker than the steel plates on other bolsters.

Stanchions must be placed in the inner sockets (where provided) of securing bolsters ("E" diagrams 2, 3 and 4).

Two chains must be used at each securing bolster ("C" diagrams 2, 3 and 4).

One chain must commence at one end of the bolster and the other chain at the opposite end, passed round the load one chain each side of bolster and secured at the opposite end from which started, leaving the tightening screws on opposite sides.

To prevent spreading, such loads must in all cases be chained at suitable points ("D" diagrams 2, 3 and 4), and the chains used for this purpose must be "free," i.e., must not be attached to the vehicle or its fittings in any way.

**Width of Loads.**

The width of each load must not exceed 2 feet 8 inches.

**Coupling.**

Each set must be coupled so as to limit the maximum possible movement between the two wagons to 1 foot 4 inches.

# LMS ACCIDENT REPORT
## Plaistow – 19th September 1927

A down freight train was leaving Plaistow sidings and travelling on to the down through line under 'clear signals' when a down passenger train, Fenchurch Street to Southend, came into sidelong collision with it at the converging junction. Eleven passengers suffered from minor injuries or shock.

Signalman Gill, of West Ham, had accepted the passenger train from Upper Abbey Mills Junction at 9.18 a.m. and received the entering-section signal at 9.21 a.m. It could not be accepted by signalman Smith at Plaistow when it was offered, the junction then being set for the departure of the freight. Gill suggested that he lowered the West Ham home signal at 9.21½ a.m., half a minute after the train had left Upper Abbey Mills Junction and when it was nearing the signal, in conformity with regulation 40. But in view of his own evidence and that of driver Petchey, and having regard to the practice, which appears to have grown up, of not checking trains upon the rising gradient of 1 in 100 when approaching this station in such circumstances, it is apparent that the rule on this occasion was not complied with. Gill was occupied watching the departure of the freight and said he did not observe the approach of the passenger.

Driver Petchey said he knew that his next stop was at Plaistow. He was driving from the left-hand side of the footplate in the direction of travel, bunker leading. He observed the West Ham distant in the 'warning' position, the Upper Abbey Mills Junction home being 'clear'. On passing the iron bridge he then observed the West Ham home at 'danger'; but after accordingly closing the regulator upon reaching the distant, he turned round, looked up again, and saw that the signal had cleared. He was not checked and had no occasion to apply the brake; but he did not, apparently, though travelling on the rising gradient, again apply steam until reaching a point some 20 yards in rear of the home. Upon passing this signal he observed the Plaistow starting signal in the 'clear' position for the freight train and kept his engine in steam to a point about half-way down the platform, where his view of the signal was then lost – not having lasted for more than 20 seconds, assuming a speed of 20 miles per hour. Evidently he then permitted himself to jump to the false conclusion that he had a clear road into Plaistow, and thereafter simply failed to observe the two intervening signals, in other words the West Ham starting signal and the Plaistow home.

Petchey is a man with an excellent record, and frankly accepted full responsibility for his failure on this occasion. He could only account for it by suggesting that over-anxiety made him pay particular attention to the Plaistow starting signal with the result that, assuming the road to be clear, he "never thought of the intervening signals," notwithstanding the fact that he was preparing to stop at Plaistow. His attention was not diverted in any way, nor was he conversing with his fireman.

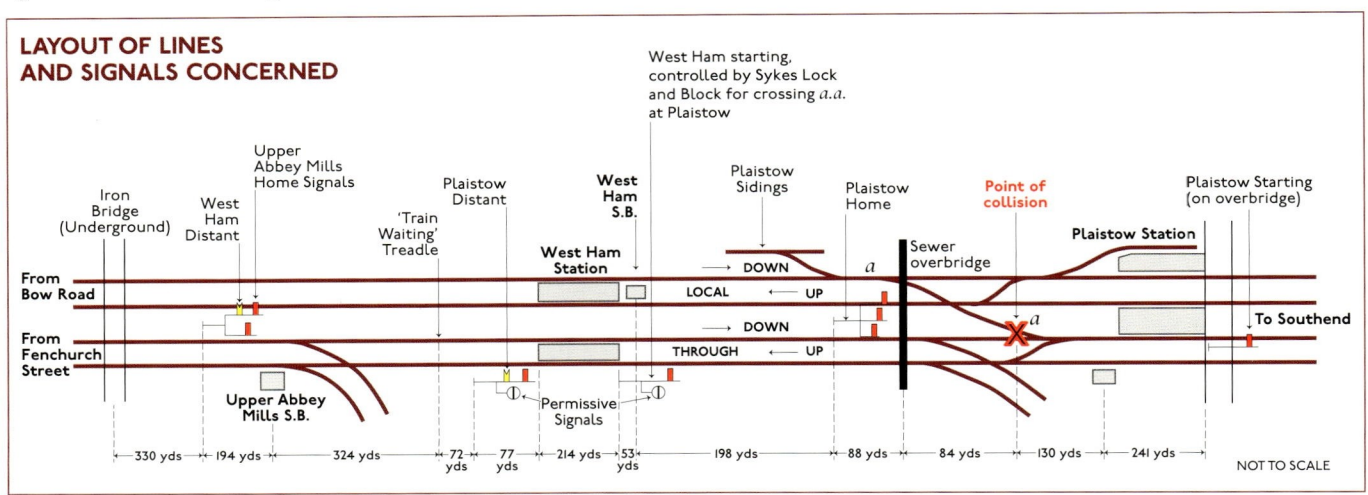

Colonel Mount remarks that it is difficult therefore to point definitely to any reason or contributory cause for his temporary lapse; but he feels that had the Plaistow starting signal been located a little lower, thus rendering it invisible from West Ham; had the Plaistow home been perhaps a little higher; or had the West Ham starting signal been carried on a bracket over the road, this accident might not have happened. Or again, had signalman Gill waited for the operation of his indicator before lowering the West Ham home, thus ensuring strict compliance with rule 40, Petchey's mind would have become concentrated upon the signals immediately ahead of him. While, therefore, Petchey cannot be relieved of the full measure of blame, it is suggested by the inspecting officer that consideration be given (a) to the practicability of improvement in the sighting of the signals concerned; and (b) to the application at West Ham of rule 40, having regard to the limited view of approaching trains which is obtainable from the box in certain circumstances, and notwithstanding the rising gradient and the situation of the junction in rear.

The report concludes by observing that the case is also illustrative of collision preventable by a system of automatic train control. On the other hand, it is probable that the provision of the less costly detonator-placing apparatus, such as is commonly provided nowadays for operation from the box, would have afforded the necessary warning; while it is difficult to believe that the indications concerned would have been disregarded had a type of colour light, instead of semaphore, signal existed, such as those recently installed upon the local lines alongside.

Standard Class 4MT 2-6-4T No. 80096 passes West Ham signal box on the approach to West Ham station with a Southend-Fenchurch Street train on 25th March 1956. The sewer overbridge highlighted in the track diagram is above the last coach of the train with the crash site just beyond.
PHOTO: A. E. BENNETT (AEB1088) © THE TRANSPORT TREASURY

LEFT: Abbey Mills Junction signal box pictured on 11th March 1978. In the distance an LT District Line train can be seen on the iron bridge behind the pylon.
PHOTO: G. H. TAYLOR (GHT5884) © THE TRANSPORT TREASURY

ABOVE: Ex-Midland Railway Class 1P 0-4-4T No. 1287 (BR No. 58043) pictured at Plaistow shed in June 1934. © THE TRANSPORT TREASURY

*Report and track diagram reproduced from The Railway Engineer, September 1928.*

# NORTH BY NORTH WEST
## ALL PHOTOGRAPHS AND WORDS BY JEFFERY GRAYER

AS EXCITING AS A HITCHCOCK THRILLER, JEFFERY RECALLS A PILGRIMAGE HE MADE 55 YEARS AGO, IN JUNE 1968, TO THE NORTH WEST OF ENGLAND TO WITNESS THE LAST WEEKS OF BR STEAM.

Rose Grove shed was one of the last quartet of Motive Power Depots to remain open to steam until the end came in August 1968. Here 'Black Five' No. 45096 rests outside the shed awaiting its next turn.

Having collected my Midland Region Rover ticket from my home station on the Sussex coast on Wednesday 19th June 1968, I made my way the following day to London's Victoria station and then to Euston to begin a week's tour of BR's remaining steam bastion in the North West of England.

The final scheduled steam trains were due to run only six weeks later, culminating in the famous '15 Guinea Special' of 11th August which signalled the end of BR steam.

I knew that, following the introduction of the summer timetable on 6th May, a further four sheds had closed to steam in Lancashire at Edge Hill, Speke Junction, Stockport Edgeley and Heaton Mersey, leaving just a handful of depots open. In the Manchester Division the survivors were:

| Newton Heath | 9D |
| Patricroft | 9H |
| Bolton | 9K |

And, in the Preston Division they were:

| Carnforth | 10A |
| Lostock Hall | 10D |
| Rose Grove | 10F |

The great variety of locomotive designs formerly operating in the area had been reduced to just seven types – the Stanier 'Black Five' 4-6-0s (the most numerous with 85 of the 842 examples constructed still on the books in mid-June), the Stanier 8F 2-8-0s with 68, BR Standard Class 5 4-6-0s (73xxx) and Class 4 4-6-0s (75xxx) with seven and five respectively, together with the sole remaining 'Britannia' class Pacific No. 70013 *Oliver Cromwell* retained for special duties plus Ivatt 2-6-0 No. 43106 which was destined for preservation on the Severn Valley Railway.

Although the last trio of 9F 2-10-0s, Nos. 92077, 92160 and 92167, were not recorded as withdrawn until the four-week period ending 13th July, I did not see any of them working during my visit. They had latterly been employed on Heysham–Leeds petrol trains, but from mid-June onwards they were to be found stored out of use at Carnforth.

Before June was out, rostered steam from Newton Heath and Bolton was to cease together with Patricroft on 1st July, leaving just three depots operating in the final few weeks.

I was travelling via Crewe, courtesy of an AL6 electric-hauled express from London, where one of the English Electric Type 4 (Class 40) locos took over for the run northwards, this being the limit of electric traction at this time.

I changed at Lancaster onto a Barrow-bound service and was duly deposited at Carnforth, enabling me to make my first shed visit that same afternoon. At the depot, in addition to numerous 'Black Fives' and 8Fs on shed roads, I found a number of stored 9Fs together with four locomotives destined for preservation in the shape of Fairburn 2-6-4 tanks Nos. 42073 and 42085, B1 4-6-0 No. 61306 and Ivatt Class 2 2-6-0 No. 46441, which were parked on a section of track rented by the Lakeside Railway Society.

I spent the next few days visiting stations in Liverpool, Preston and Manchester and the sheds at Newton Heath, Lostock Hall, Patricroft and Rose Grove.

Regular steam-hauled passenger turns were by then pretty thin on the ground with steam haulage of the Manchester Victoria–Heysham Harbour 'Belfast Boat Express' having finished on 5th May when 'Black Five' No. 45025 had taken out the last working.

If you were "in the know" however, you could still travel by steam; the 20.50 Preston–Blackpool South and 21.25 Preston–Liverpool, for example, continued to be steam hauled until Saturday 3rd August when 'Black Fives' Nos. 45212 and 45318 respectively did the final honours.

The following day saw what was probably the last 'normal service' steam working in the shape of No. 45212 shunting the sleeping cars from the 23.45 Euston–Preston into the bay at Preston, although the passengers were probably not sufficiently awake to realise the significance of the moment!

**With a feed pipe disappearing out of the picture to another watering bag, 'Black Five' No. 45104 takes water from the tower adjacent to Manchester Exchange's famous long platform.**

During the author's visit, Newton Heath shed, as with all sheds at this time, provided no problem gaining access. Although many of the remaining locomotives would obviously not be working again, there was still enough live steam, including 'Black Fives' Nos. 44890 and 45202, to provide the sights, smells and sounds of an active depot.

4th August also saw the running of six special trains touring Lancashire, the following day seeing the formal termination of all normal steam services. There had been doubts about this final steam date right up to the last minute as there were concerns that deliveries of the last of the English Electric Type 4s could not be made in time.

The grapevine had it that even if there were deficiencies in the supply of Type 4s then pairs of Type 2s would be operated instead, or the loads on certain freight services would be reduced to within Type 2 limits, to ensure the demise of steam on schedule. In the event there were still three of the 50 Type 4s outstanding at the beginning of September.

The dour Manchester Victoria of the late 1960s, coated in decades of soot-encrusted grime and in a semi-derelict condition over large parts of the complex, could still boast the 'Platting Bankers' on standby to assist freight up the 1-in-47/1-in-59 incline to Miles Platting.

I was able to photograph a number of 'Black Fives', including Nos. 44780, 45255, 45206 and 45330, on these duties. Steam, in the shape of more 'Black Fives', Nos. 45104, 45203 and 45394, could also be seen from the 2,914ft platform, the longest in the country at the time, connecting Victoria and Exchange stations. Even then Exchange was under sentence and was to close the following May.

I took the short ride from Victoria, on an Oldham-bound DMU to Dean Lane station, from where it was only a short walk to Newton Heath shed, the station at Newton Heath itself having closed in January 1966.

As with all sheds at this time there was no problem gaining access and although many of the remaining locomotives would obviously not be working again, there was still enough live steam, including 'Black Fives' Nos. 44890 and 45206 on the day of my visit, to provide the sights, smells and sounds of an active depot.

Although Lostock Hall continued to provide steam power for the Colne–Preston and return parcels services throughout July, a visit to the shed was pretty depressing as the yard contained row upon row of condemned engines and very little in steam.

However, Preston station could still produce a number of steam-hauled freights and I particularly remember BR Standard Class 4 4-6-0 No. 75027, preserved at the Bluebell Railway currently awaiting overhaul, running through from the former Blackburn line with a ballast train. No. 75027 had latterly been used on freight trains on the Grassington branch which closed in June, and had been maintained in immaculate condition by local enthusiasts.

Seemingly on a collision course, a pair of 8Fs are in fact following each other through Rose Grove station, having just left the adjacent shed which is behind the photographer.

Patricroft still boasted several Standard Class 5 4-6-0s and one of the quartet of Caprotti-valve versions then remaining in traffic, No. 73125, was just in the right position on the morning of my visit for a photograph in front of the shed with No. 73050, which was later bought direct from BR and preserved as *City of Peterborough*, lurking in the shed behind.

An 8F, No. 48491, was captured under the mighty ferro-concrete coaling stage, which then dominated the site. The following day I took a trip out to Rose Grove near Burnley where No. 75027 was seen again, being turned on the turntable, keeping company with a number of 'Black Fives'.

This was to prove my last ever visit to an active BR steam shed. By this time photographers were easily outnumbering shed staff and all the old restrictions about visiting such installations were forgotten, in the scramble to record the death throes of the steam age.

Unfortunately I never did get to Bolton shed, only managing to photograph the motive power depot and yard from a passing train.

I rounded off my sojourn in the north west by taking a trip on the Settle and Carlisle line on 25th June before returning to London the next day, unfortunately during an overtime ban and period of 'Work to Rule' by NUR and ASLEF staff, which memorably left me stranded in the capital in the early evening with no more trains that day!

Still such inconveniences were as nothing in comparison with the week of steam that I had just experienced and which of course could never be repeated.

RIGHT: Erupting from Manchester Victoria station in a volley of noise, steam and smoke, is an unidentified 'Black Five', epitomising the power and glory of the steam locomotive; a sight that would be gone from the network forever in a matter of weeks.

OPPOSITE PAGE: Lostock Hall presents the sight of lines of condemned locomotives awaiting their final journey.

| Withdrawal of steam from the North West ||||
|---|---|---|---:|
| Week period ending | Locomotive Class | Locomotives withdrawn | Total |
| 15th June 1968 | LMS 'Black Five' 4-6-0 | 44803, 45187, 45345, 45411 | 4 |
| | LMS 8F 2-8-0 | 48267, 48282, 48327, 48338, 48374, 48380, 48384, 48467, 48549, 48646, 48687, 48746 | 12 |
| | BR Standard 9F 2-10-0 | 92091, 92118 | 2 |
| | | | Period Total: 18 |
| 13th July 1968 | LMS Ivatt 4MT 2-6-0 | 43106 | 1 |
| | LMS 'Black Five' 4-6-0 | 44777, 44780, 44802, 44818, 44845, 44878, 44884, 44890, 44891, 44910, 44929, 44942, 44947, 44949, 45046, 45076, 45104, 45149, 45202, 45203, 45209, 45255, 45290, 45312, 45382, 45420, 45435, 45445 | 28 |
| | LMS 8F 2-8-0 | 48026, 48033, 48115, 48132, 48168, 48170, 48212, 48293, 48319, 48321, 48323, 48356, 48368, 48369, 48373, 48392, 48491, 48504, 48529, 48546, 48612, 48620, 48652, 48678, 48692, 48720. | 26 |
| | BR Standard 5MT 4-6-0 | 73010, 73050, 73125, 73133, 73134, 73143 | 6 |
| | BR Standard 9F 2-10-0 | 92077, 92160, 92167 | 3 |
| | | | Period Total: 64 |
| 10th August 1968 | LMS 'Black Five' 4-6-0 | 44690, 44709, 44713, 44735, 44758, 44806, 44809, 44816, 44874, 44877, 44888, 44894, 44897, 44899, 44932, 44950, 44963, 44971, 45017, 45025, 45055, 45073, 45095, 45096, 45134, 45156, 45200, 45206, 45212, 45231, 45260, 45262, 45268, 45269, 45287, 45305, 45310, 45318, 45230, 45342, 45350, 45353, 45386, 45388, 45390, 45394, 45397, 45407, 45444, 45447 | 50 |
| | LMS 8F 2-8-0 | 48062, 48167, 48191, 48247, 48253, 48257, 48278, 48294, 48340, 48348, 48393, 48400, 48410, 48423, 48448, 48476, 48493, 48519, 48665, 48666, 48715, 48723, 48727, 48730, 48752, 48765, 48773, 48775 | 28 |
| | BR Standard 5MT 4-6-0 | 73069 | 1 |
| | BR Standard 4MT 4-6-0 | 75009, 75019, 75020, 75027, 75048 | 5 |
| | | | Period Total: 84 |
| 7th September 1968 | LMS 'Black Five' 4-6-0 | 44781, 44871, 45110 | 3 |
| | BR Standard 7MT 4-6-2 | 70013 | 1 |
| | | | Period Total: 4 |

Stanier 'Black Five' 4-6-0 No. 45055 waits in the parcels bay at Manchester Exchange station having delivered a short van train. There seems to be plenty of platform trolleys to help unload the contents.

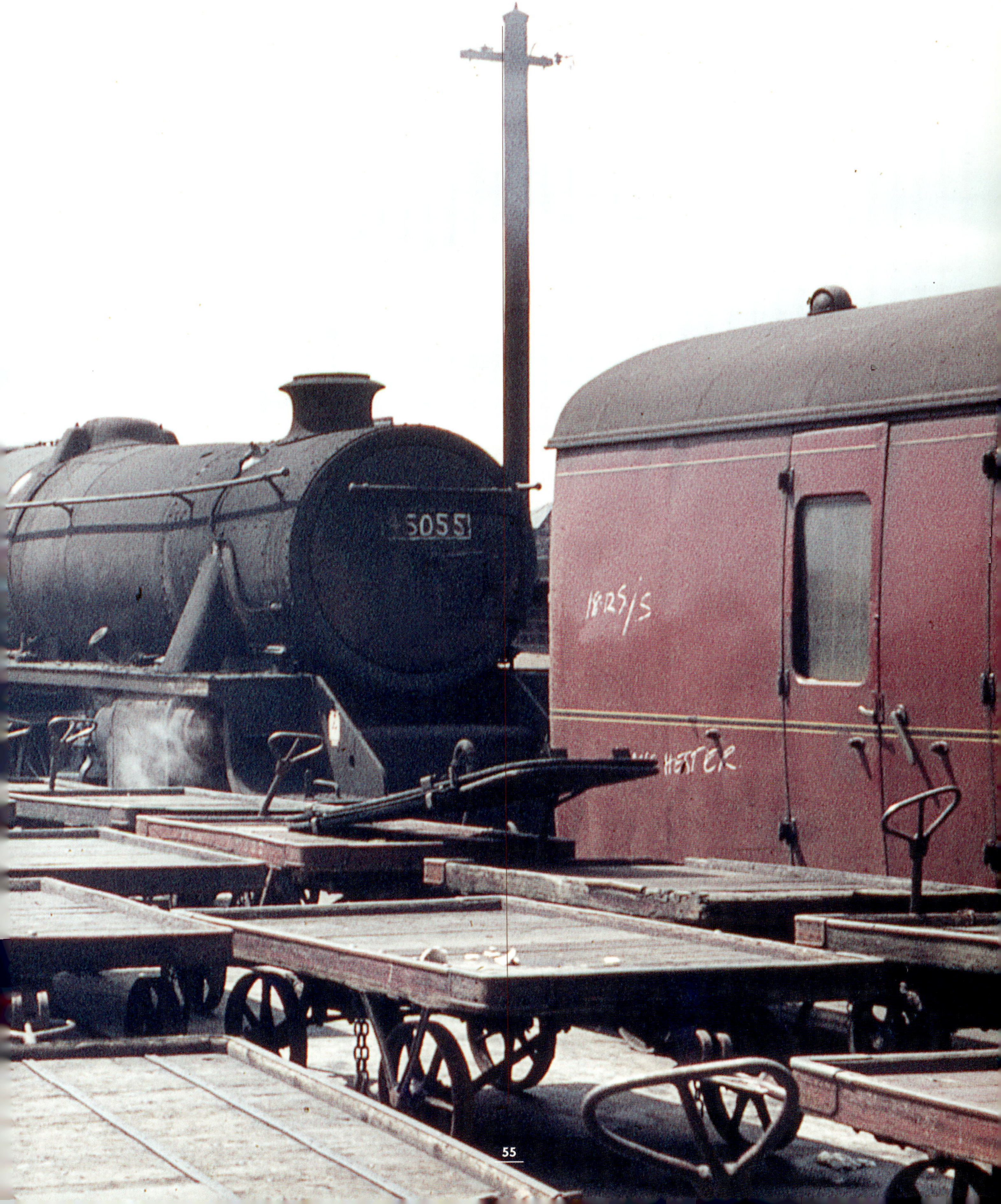

## FURY – THE LMS HIGH-PRESSURE EXPERIMENT
### by David Cullen

"There is no doubt that there is a general tendency through the locomotive world to move in the direction of much higher steam pressures, and I am of the opinion that if the proposal is looked upon in the lines of research, it would be advantageous for us to make a trial of some type of high pressure locomotive."

This recommendation was made to the board of the London, Midland and Scottish Railway by Mr. Henry Fowler, Chief Mechanical Engineer. 'Advantageous' in Fowler's report referred to the economics of such an engine. High-pressure steam is used in shipping boilers and power stations. Having special properties through which it effectively generates its own heat, this results in a lower rate of fuel requirement compared with regular boilers.

Fowler's intention was to cut coal use through the adoption of such a boiler on LMS locomotives. Eager to explore the feasibility, the board approved a project for an experimental locomotive using steam at pressures never before contemplated. Due to its advanced design, the construction cost was estimated at £8,750; a £1,000 increase over any previous LMS machine.

It was expected though, that this would be offset by a £130-£140 annual reduction in fuel costs. If the prototype proved successful, this would be further multiplied by the number of additional engines built.

The locomotive resulted from a collaboration between the LMS, The Superheater Co. Ltd., largely responsible for design work, and Glasgow-based North British Locomotive Co. Ltd., for its construction. The latter arrangement stemmed from the LMS simply not having the facilities to produce anything of this size and complexity. Built in 1929, it was given the running number 6399 and the name *Fury*, originally carried by a Fowler 'Royal Scot' class 4-6-0. This had relinquished the name to take that of a British army regiment. *Fury's* outline bore a close resemblance to the 'Royal Scots', although certain differences were apparent, including external recesses accommodating sundry equipment, and the firebox; built to the maximum height permitted by the LMS loading gauge; its top surface extending forward over the boiler some 10ft beyond the flat side panels.

*An interesting 'head on' view of FURY taken as part of the short sequence of official photographs at Hyde Park Works.*

*Fury* was given a water-tube boiler based upon the 'Schmidt-Henschel' concept; a highly complex piece of engineering compared with a regular fire-tube boiler. This had been pioneered to advance German steam, which, using conventional technology, had reached its peak within the confines of the country's loading gauge. 'Bigger' was no longer an option; 'alternative' had to be explored.

*Fury's* boiler comprised three sections; each using an individual process to generate its own specific pressure. The first, located just ahead of the cab, comprised an arrangement of vertical water-tubes forming a self-contained, closed circuit. This was charged with distilled water to avoid scale build-up.

The lower ends of the tubes were set in a foundation ring and the base of the combustion chamber. Forming the walls of the firebox, the tubes would be subjected to immensely high temperatures, producing within them steam at 1,400-1,800 psi, and at around 350 degrees centigrade, depending on the intensity

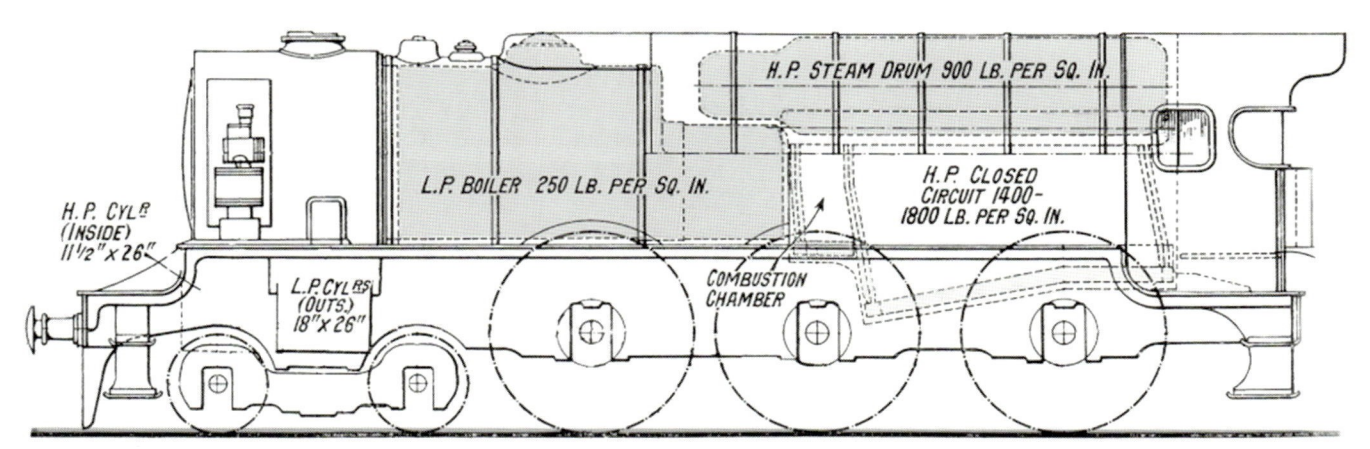

The arrangement of the three main Schmidt boiler compartments shown within the constraints of a Royal Scot locomotive outline.

of the fire. So potentially hazardous was this, it has since been queried how it ever gained approval in the first place.

This ultra-pressure steam was not used to power the locomotive. Instead, it provided a heating medium for generating the operational steam within the second section located above. Here, the upper ends of the tubes expanded into two cylindrical equalising drums, some 12ft. in length by just over 13in. diameter, one located horizontally on each side. From these, coils carried the ultra-pressure steam to evaporation elements set within a third, much larger drum. Supplied by the specialist boiler firm John Brown & Co. Ltd. of Sheffield, this was a machine-forged unit 15ft. 2in. in length by 3ft. 7½in. diameter. Never in direct contact with the fire, this was constructed of nickel-steel. It occupied the space above the equalising drums; its lower portion between their tops. The elements turned water in this drum into operational steam at 900 psi, unprecedented in this country and double that used in another high pressure locomotive being tried at the same time by the London & North Eastern Railway – Gresley's 'hush-hush'.

The third section comprised a regular locomotive boiler, which together with its smokebox formed 6399's front half. Measuring 5ft. 7½in. by just over 13ft in length, the boiler had a nickel-steel barrel with mild steel front and rear tube-plates and normal fire-tubes. Steam was generated within this at a comparatively modest 250 psi. A pump supplied the high-pressure drum with water from this boiler which was itself fed by live and exhaust steam injectors fitted on the driver's and firemen's sides respectively. Steam was generated more rapidly in the HP drum than in the low-pressure boiler, so a special valve was fitted to transfer any excess in the former to the latter, rather than let it be wasted through the safety valves.

To make optimum use of its HP system, *Fury* was built as a 'compound' locomotive. Used extensively on the Continent and elsewhere, only a very few different types of compound were built in this country. In the compounding process, steam was used twice; firstly at boiler pressure to power one or more cylinders, then having lost some pressure, it was fed to operate low-pressure cylinder(s) before discharge from the chimney. No. 6399 operated on a single HP cylinder at 11½in. in diameter by 26in. stroke. Set below the smokebox between the main frames, this drove the engine via a crank web in the leading coupled axle. Two LP cylinders, each 18in. by 26in., were located externally and connected to normal crank-pins on the middle coupled wheels. Each cylinder was fitted with an individual set of Walschaerts valve gear.

Just as the boiler was complex compared with that of a regular locomotive, so was the process through which the steam was utilised. The regulator handle operated both high and low pressure valves. When opened, 900 psi steam from the large drum passed via a superheater to operate the single HP cylinder. Simultaneously, the 250 psi steam was released from the conventional boiler, although not flowing straight to the LP cylinders, but into a special mixing chamber. Where it would combine with the exhaust steam from the HP cylinder. This 'mixture' then passed through an LP superheater to power the external cylinders before final discharge.

The boiler tubes gave an evaporative surface of 1,335sq.ft. A further 218sq.ft were provided by the firebox, which contained a grate area of 28 sq.ft. The HP superheater elements provided 274 sq.ft of heating surface. These were located in the lower boiler tubes. The LP elements gave 355 sq.ft and were fitted in the upper tubes. *Fury* was 64ft. 3in. long over the buffers and just over 13ft. 2in. from rail to chimney crown. The six coupled wheels were 6ft. 9in. in diameter and the bogie wheels 3ft. 3½in. In full working order, No. 6399 weighed 130 tons 6 cwt; the locomotive 87 tons 2 cwt and the tender 43 tons 14 cwt. Weight of adhesion was reckoned at 63 tons 2 cwt and tractive effort calculated as 33,200lb. Set on six, 4ft. 3in. diameter wheels, the tender carried 5½ tons of coal and 3,500 gallons of water; smaller quantities than a 'Royal Scot' tender due to the foreseen economies of the high-pressure system. In appearance, it was old-fashioned and far less impressive than the true 'Scot' version.

The locomotive was presented to the LMS in December 1929. Scheduled for work in Scotland, it was sent to Polmadie depot in Glasgow on Thursday 6th February 1930. Unfortunately it was to have the shortest operational period of any locomotive before or since, and was never to head a revenue-earning train. It was withdrawn just four days later, following an outing to run-in its wheel bearings, when its ultra-high pressure system led to the name *Fury* being tragically appropriate. Four men were on the footplate that day: Driver Hall, Fireman Blair, Inspector Louis Schofield from designers The Superheater Co. Ltd., and a Mr. Pepper, another inspectorate official.

Passing Carstairs station in Lanarkshire, a jet of superheated steam laden with burning coals erupted from the firebox. Standing directly in its path, Inspector Schofield received terrible scalds and burns. Fireman Blair sustained scalding to an arm, followed by cuts and abrasions after leaping from the footplate. Inspector Pepper used a handrail to hang outside the cab, avoiding the worst. At his driver's station to the cab side, Mr. Hall was also relatively unscathed. The blast having subsided, when showing great courage and professionalism, he brought *Fury* to an emergency stop. Medical assistance was summoned and the men were all taken to the Glasgow Royal Infirmary. The most seriously injured, Inspector Schofield, died the following day. Fireman Blair was treated for scalds, serious but not life threatening. Driver Hall and Inspector Pepper were treated for shock and minor heat-related injuries. Both made good recoveries. Some 30 years after the incident, Mr. Pepper was made Assistant to the CME of British Railways.

*Fury* was taken out of service and an investigation ordered. The engine was duly stripped down and examined. The circumstances of the incident were deemed as follows: A 5in. longitudinal split in an ultra-high pressure tube of the closed-circuit had caused the eruption. Failure of natural circulation within the tubes had led to the tube weakening through local overheating. This had probably been due to a flaw in the metal, so minute as to have avoided detection by even the most stringent examination. This tube was a component in a sealed system of modest volume. Its contents had thus discharged rapidly. The other two sections of the boiler were not compromised; their water and steam remaining safely contained. Accordingly, though devastating, the eruption had been brief, enabling Driver Hall to quickly bring the engine to a halt.

The damage was rectified and further operational testing commenced in July 1932, but Carstairs was still a terrible memory, and the project's cancellation was ordered soon after. Following a final test run on 14th February 1933, No. 6399 was quietly despatched to the LMS workshops at Derby. Here it languished until 1935, when Mr. William Stanier, the railway's new CME, scrapped the Schmidt boiler and compound cylinders, and used the salvaged frames to begin construction of a brand new locomotive. Given the name *British Legion* and the running number 6170, this was the first of the rebuilt 'Royal Scot' class with the new tapered boiler and of conventional simple-expansion type.

Having invested some £20,000 in *Fury*, the Superheater Co. Ltd., bore the greatest financial loss from the accident. They attempted to retrieve this, or at least a reasonable percentage, but Henry Fowler had included in their original agreement a clause restricting LMS liability to £1,500. Furthermore, this was only payable if the locomotive "was found to be satisfactory in its service". It was successfully argued this was not the case. Following lengthy negotiations, an *ex-gratia* payment of £3,000 was agreed; this being paid to the company in July 1934.

From a cost point of view, the use of high pressure steam in locomotives was a sound idea, and as stated, *Fury* was not the only such experiment in this country. However, theory has the tendency to be far removed from practical reality, and it was never to prove the success hoped for. It may have had modest success cutting fuel consumption, but this is just one aspect of railway operation. Using highly complex equipment, these locomotives required specialist maintenance; more complicated and expensive than with regular ones. Revenue loss during such services would have been disproportionately excessive. These factors tended to cancel out, even outweigh any fuel saving.

Further, experiments involve things untried. In steam locomotives, this tended to result in a high degree of mechanical failure, with attendant needs for investigation, repair, modification and testing, in attempts to 'get it right'. In the case of *Fury*, the Carstairs incident took this far beyond inconvenience. Although further advances were made across Britain's railway companies, and subsequently under British Railways, the Schmidt-boilered No. 6399 remained a one-off on the LMS. For reasons both economic and moral, no such experiment was ever again carried out.

FURY, newly completed at the Hyde Park Works of the North British Locomotive Co. Ltd. A dozen workmen pretend to do things for the official photographer.

Leaving the works, following which, many more static photographs were taken in addition to the short Gaumont British Pathe film.

# REMINISCENCES OF THE LMS AND MacBRAYNES
## Peter Tatlow, President of the LMS Society

What you may ask does the LMS have in common with MacBraynes shipping activities off the west coast of Scotland? Quite a lot as I hope to demonstrate. They both provided a public transport service to widely scattered small communities in the Western Isles of Scotland and the lightly populated Highland countryside. In both cases, a similar range of problems faced both national and local governments in trying to satisfy the demands of the communities, often vociferous, being served, while endeavouring to justify even the minimum of grants to the wider electorate. Furthermore, not only did the company meet and exchange traffic at rail heads such as Oban, Mallaig and Strome Ferry/Kyle of Lochalsh, but as we shall see, the LMS was to find itself owners of half of MacBraynes!

Steam ship operations began on the River Clyde in the first half of the 19th century leading to the development of a network of sailings from Glasgow, out along the Firth of Clyde and up the west coast of Scotland. In the early days, the forerunners of MacBraynes soon found a happy medium whereby the highly popular summer cruises could subsidise lightly trafficked more remote destinations, especially during the winter months, helped by the grants earned for carrying the Royal Mail. This worked well enough while the tourists continued to come; the outbreak of war with Germany in August 1914, however, left the company to shoulder the burden of maintaining the loss-making services to provide for islanders with no profit from elsewhere with which to make up the deficiency. Despite intermittent bailouts, matters did not improve much after the war and with an ageing fleet and one or two losses at sea, in 1928 MacBraynes indicated to the Government that they could no longer continue to function and would have to withdraw from the Mail Contract.

At this point, the Government looked around and turned to the shipping company Coast Lines and the LMS for help which culminated in the setting up of David MacBrayne (1928) Ltd. Hereafter replacement ships, diesel rather than steam propelled, regularly appeared until conflict with the Germans again caused severe disruption to activities during World War 2. Subsequently the prospect was nationalisation of the railways from 1st January 1948 (which will account for railway staff being able use privilege tickets on MacBraynes). Up until the 1960s, the occasional motor car had been catered for either by hoisting them aboard while supported by a sling around each wheel, or if tides were remotely suitable, driving precariously over a pair of sometimes very inclined wooden planks spanning between the ship's gunwale and the quayside.

The ubiquitous Stanier Black Five No. 5362 takes water at Balquhidder while heading a Down express passenger train to Oban.
Photo: O. S. Nock – author's collection

Soon after leaving Balquhidder, the train is faced with climbing hard as it threads the narrow Glen Ogle while partly supported by a masonry viaduct on side-long ground on its way to the summit of 941 feet (AOD). It was a rock fall in the vicinity that led on 27th September 1965 to the premature closure of this section of line.
PHOTO: UNATTRIBUTED, AUTHOR'S COLLECTION.

MacIntosh 4-6-0 Class 191 No. 14624 free wheels its six-coach train down the bank from Glencruitten on the approach into Oban in April 1925.
PHOTO: KEN NUNN – AUTHOR'S COLLECTION

During the above mentioned two World Wars, one needed a permit to enter much of the North of Scotland and the Western and North Isles. These were available only to residents and those with duties there. Following VE Day (Victory in Europe – 8th May 1945) these restrictions were lifted and my parents decided to seize the opportunity during late August/early September to take my brother and me on a two-week holiday to Dervaig on the Isle of Mull. We travelled north on 25th August 1945 by the 7.30pm train from Euston bound for Inverness and Aberdeen, but part of which ended up at Oban. Only on Fridays, however, did this last portion include a composite sleeping car in which the family occupied a four-berth third class compartment, but as yet the train included no dining car. The Oban section was dropped off at Stirling shortly before six in the morning, where, being summertime and fully daylight, my father left the train and set off into the town to see if he could procure something for breakfast. In the meantime, while he was away, we became agitated when our coaches were moved to another platform. We need not have worried though; we were only being repositioned within the station to await the early morning trains from Edinburgh and Glasgow for Perth and Aberdeen with portions for Oban. He returned with that Scottish speciality of freshly-baked mutton pies from a baker who had only just opened for the day's business.

In due course, at a little after eight, we were coupled up to a train again and from Callander headed into the Trossachs and westwards through the mountains, where the road and railway are seldom far apart. Upon arrival at Oban station just before noon, we hired a porter to convey our luggage to MacBraynes' pier, there to board MacBrayne's *MV Lochinvar* for the sail to Salen, calling first off Craignure and at Lochaline. In those days there was no pier at Craignure at the southern end of the island

Before World War 2, ex-CR 4-6-0 Pickersgill 191 class No. 14622 leads an unidentified ex-HR 4-6-0 Clan class, with steam to spare, darkening the sky as the pair leave Oban and commence the 1 in 50 gradient for three miles to reach the summit at Glencruitten Crossing.
PHOTO: P. RANSOME-WALLIS – AUTHOR'S COLLECTION

The first deliveries of Stanier's Class 5 locomotives to the Highland Section released eight of the HR's Clan class and two of its Castle class 4-6-0 in the mid-30s to work the Callander and Oban line. No. 14768 CLAN MACKENZIE enters the loop at Strathyre, gently blowing off steam on a seven coach express from Glasgow to Oban. PHOTO: O. S. NOCK – AUTHOR'S COLLECTION

and small boats came out to meet us anchored in the bay. To see people climbing aboard was one thing, but our eyes were really opened as we watched sheep loaded by being thrown and caught by a pair of burly Highlanders. For our return two weeks later, it meant an early start to catch *Lochinvar* again on her inward run to Oban. Breakfast was taken on board.

The return working of the sleeping car for London only ran on a Monday night, which was unsuitable; I expect my father had to be back at work. So instead, we spent a night in Oban, before catching the midday train to Glasgow and overnight sleeper from there. In the interim, however, I recall watching the early evening train stretching beyond the end of the platform waiting to start for Glasgow, headed by a pair of Stanier Black Fives, both blowing off steam. Little concern seemed to be apparent at the late departure, to which I was informed that in these parts, life was regulated by the clock, but rather the comings and goings of the train provided the punctuations of the day. Once the departing train had disappeared from view, we wandered back towards the centre of the town, but still the thunderous sound of the labouring locomotives persisted and as a nine-year-old boy I wanted to see them. The noise of the engines' exhausts, as they climbed the 1 in 50 gradient in a great circle around the outskirts of the town and up to Glencruitten Crossing with a heavy train, reverberated around the town square for a good ten minutes. The next day we sampled the trip for ourselves as we took the midday train to Glasgow Buchanan Street. During the passing of a train in the opposite direction at one of the single line crossing loops, a named engine was pointed out to me and I have always wondered whether this was one of the Highland Railway's migrants – the 4-6-0s of the Clan class, but I was probably a decade late. Later the same day we caught the overnight train from Central Station to London Euston and home.

It was August 1955 before a Forces' leave warrant was to afford me the opportunity of another visit north of the Border. A round trip, starting with an overnight sleeper train, the Royal Highlander from Euston to Inverness, was planned, followed by

MV Lochnevis was one of the new tonnage placed in service during the pre-war period by the MacBrayne (1928) regime managed by the Coast Lines and the LMS following their take over. Here MV Lochnevis approaches at Kyle of Lochalsh on the mail run from Portree to Mallaig during a morning in June 1937. PHOTO: W. H. NASH, COURTESY KATE ROBINSON

MacIntosh 4-6-0 Class 55, No. 14602 prepares to return with the two-coach branch train from Ballachulish on 18th June 1925, with the exposed face of the slate quarry in the background. PHOTO: R. S. CARPENTER COLLECTION

My last sighting of MacBraynes' MV LOCHINVAR was as she sailed into Mallaig harbour on the morning of 4th September 1959 to the sound of the pipes drifting across the water, on the inward mail run from Portree with a party of guests returning from a society wedding on the Isle of Skye. The bagpiper stands against the taffrail just to the left of the funnel. This was her last season with MacBraynes before being sold out of service and sadly lost with all hands on 3rd April 1966. PHOTO: P. TATLOW (35/31-35)

a trip on the Dingwall and Skye to Kyle of Lochalsh line, with an afternoon sail on *MV Lochnevis* to Portree, MacBrayne bus to Kyleakin and ferry back to Kyle. The next morning *MV Lochnevis* took me down the Sound of Sleat to Mallaig and on by train along the West Highland Extension to Fort William. Later a MacBraynes tour was undertaken by bus to Glenfinnan, motor boat cruise down Loch Shiel, further buses and a ferry to Fort William. The next day a bus and the Ballachulish branch train were taken to Oban. Following a couple of MacBraynes' cruises, one to the sacred Isle of Iona, Fingal's Cave on Staffa and around the Isle of Mull in *TSS King George V*; and another up Loch Sunart in *MV Lochfyne* were indulged in. As sleeping cars from Oban were by now nightly, except Sundays, the 6.00pm train to Euston, again in a third class four-berth compartment, but this time shared with others, was used for the return south. One of my fellow passengers, a recently retired bachelor baker, was setting off on a cruise, an exotic thing to do in those days. He and another passenger were determined to see the night through by the consumption of more than adequate nips of whisky, which did not bode so well for the rest of us. Arrival at Euston scheduled for 8.15am was an hour late – par for the course in those days.

For the record, my last visit to Oban, during the transition from steam to diesel traction, by car, was made on 24th July 1961, when a look was made around the engine shed. Diesel locomotives had already taken over from steam on the line to Kyle of Lochalsh and pairs of North British Type 2 diesels were hauling some trains on the Oban line, the Ballachulish branch was still steam worked, but the glamour had gone.

### References

Duckworth C. L. D. & Langmuir G. L., *West Highland steamers*, Richard Tilling, 1950.

Clark A., *The Making of MacBraynes- A Scottish Monopoly Spanning Three Centuries*, Stenlake Publishing Ltd, 2022.

Tatlow P., *The Royal Highlander & sleeping car services to the North of Scotland, Backtrack*, 1998, pp S57-S63.

Tatlow P., *The Development of the railways to NW Scotland, Backtrack*, 2015, pp218-223, 357-36 & 452-457.

Tatlow P., *Locomotives for NW Scotland, Backtrack*, 2016, pp12-18.

Tatlow P., *The LMS in Scotland, Backtrack*, Vol 7, 1993, p33-41.

## THE 'KNOTTY' IN THE LATE STEAM AGE
### Words and Photos by Alan Postlethwaite

The North Staffordshire Railway was unusual for having no local rail competitor. Centred on Stoke-on-Trent, its domain extended to Macclesfield in the north, Uttoxeter in the south east and Crewe and Market Drayton in the west. I photographed its southern tentacles in 1962 while seconded for six months to Stafford. I discovered some interesting modern freight and passenger trains. The stations left a strong impression of a very proud railway.

Awaiting electrification, the 'late steam age' was also the 'age of the interim diesel'. Pictured near Trentham, English Electric Type 4 No. D218 heads an express to Euston. The LMS was fond of 'Calling-on' signals.

Not in published timetables, Wedgwood Halt was for the exclusive use of workers at the adjacent porcelain factory. Built of wood upon concrete pyramids, it had electric lamps on the Up platform and oil lamps on the Down.

Trentham station had a separate platform for the short tourist branch to Trentham Gardens. The branch closed in 1957, seen here in 1962 with assorted hopper wagons on the stub.

The clear lettering says it all on the locomotive and 20-ton private owner wagons. To serve the extensive coal sidings of Meaford power station, the Central Electricity Generating Board's 0-6-0 tank No. 1 was built in 1951 in Newcastle by Robert Stephenson & Hawthorn Ltd. Following closure of the 'A' station, this engine was bought by the East Lancashire heritage railway, hauling their first trains in 1987. It is currently preserved on the Stephenson Steam Railway at North Shields as No.1 TED GARRETT JP, DL, MP.

Coal from the North Staffordshire coalfield was carried in assorted wagons along the Trent Valley line to Meaford power station. Seen here are a full train near Trentham (ABOVE), headed by Black Five No. 44871, and empties near Wedgwood Halt (BELOW) behind LMS 'Crab' No. 42887. At the power station, a works locomotive would propel each wagon onto a tippler which rotated sideways through 360 degrees. Starting in the late 1960s, the final batch of CEGB inland coal-fired power stations would be served by 'fitted' merry-go-round trains. These were diesel-hauled with bottom-hopper wagons which were unloaded at slow speed without uncoupling.

The Trent Valley main line of the 'Knotty' split at Stone Junction, veering right to Stafford and Birmingham and straight on to Rugby and Euston. Stone station was V-shaped with the wooden signal box at the apex, a North Staffordshire Type 2. The 2-6-4 tank is No. 42509 and there is another engine blowing off on a goods train.

ABOVE: Stone station building was at the end of a long approach road in the 'V' between the two railway routes. The right-hand fence was truncated to allow direct loading of goods. Beyond the running lines on the right is the goods shed and a new shed. Platforms on the London side have been demolished and a concrete footbridge serves the Down platform from Stafford. The drying kilns are our first glimpse of the Potteries.

LEFT: Stone had a brick building in Jacobean style with round-topped windows and an entrance portico. The Flemish gables are finished with stone copings, finials and coats of arms. The scene has a busy air with two railwaymen, two ladders and an open door to the parcels office. Let us buy a ticket!

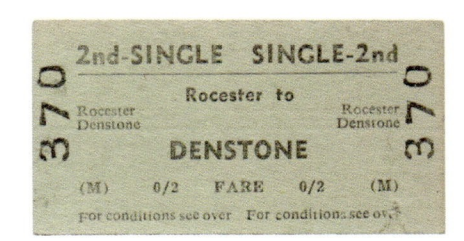

ABOVE: Near Denstone on the Churnet Valley line from Leek and Macclesfield, class 4MT tank No. 42593 heads a train to Uttoxeter on a transition between cutting and embankment.

BELOW: At Uttoxeter, class 4MT tank No. 42609 arrives from Leek. At this junction, lines led to Derby (GNR), Nottingham (MR), Burton-on-Trent (NSR), Stafford (GNR), Stoke-on-Trent (NSR), Leek (NSR) and Buxton (LNWR).

INTO MY HEART AN AIR THAT KILLS
FROM YON FAR VALLEY BLOWS:
WHAT ARE THOSE DARK REMEMBERED HILLS,
WHAT TRAINS, WHAT FARMS ARE THOSE?

THAT IS THE LAND OF LOST CONTENT,
I SEE IT SHINING PLAIN,
THE LITTLE RAILWAYS WHERE I WENT
AND CANNOT COME AGAIN.

With apologies to A.E. Housman who wrote 'Blue Remembered Hills'.

BELOW: The train sits nicely in the landscape near Rocester,

ABOVE: North of Uttoxeter, class 4MT tank No. 42663 heads a train of hoppers and box vans.

BELOW: Winter trees and an unidentified class 4MT tank near Denstone in the Churnet valley. Closed under the Beeching axe, a section of this line is preserved near Leek as the Churnet Valley Railway, running from Kingsley and Froghall to Ipstones.

Two shots of class 4MT tank No. 42609 passing through the (closed) Norbury station,. This picturesque branch led up the Dove Valley to Ashbourne where it made an end-on junction with an LNWR branch to Buxton.

ABOVE: Barlaston and Tittensor was the first station north of Stone. Its staggered platforms were beautifully kept.

BELOW: Mow Cop and Scholar Green was just north of the closely packed rail network of Stoke-on-Trent. The flower beds are a delight and the station building is extraordinarily large. This was taken on my only venture to the city, made primarily to visit Arnold Bennett's home, now a museum. Many of his novels are set in the Five Towns.

## THE PLATFORM END

Picture: Garsdale station looking north towards Dandry Mire and Moorcroft viaduct.

In future issues our aim is to bring you many differing articles about the LMS, its constituent companies and the London Midland Region of British Railways. We hope to have gone some way to achieving that in this issue.

Midland Times welcomes constructive comment from readers either by way of additional information on subjects already published or suggestions for new topics that you would like to see addressed. The size and diversity of the LMS, due to it being comprised of many different companies, each with their differing ways of operating, shows the complexity of the subject and we will endeavour to be as accurate as possible but would appreciate any comments to the contrary.

We want to use these final pages as your platform for comment and discussion, so please feel free to send your comments to: midlandtimes1884@gmail.com or write to:
Midland Times, Transport Treasury Publishing Ltd., 16 Highworth Close, High Wycombe HP13 7PJ.

# THE PLATFORM END – READER'S LETTERS

Dear Sir,

Further to your Patriot article, I attach a photo of my No. 5521 RHYL nameplate.

It seems that the crests were added following its rebuild. The other nameplate is fixed to the wall of the Rhyl Council offices but without a crest, the inscription on the plate translates to 'A fair haven on the crest of a wave'. I have seen many photos of the engine in its original condition; none have ever carried the crest.

I chased up the Rhyl nameplate because I was born there in 1931. I bought the plate minus crest from a Derby fellow who advertised it. The crest was found in a garage near Wigan and cost me £75. I trust this information is of use to you.

**KIND REGARDS**
**GRAHAM WARBURTON**

Hi Peter

I enjoyed the photograph of 'The Irish Mail' on page 63 of Midland Times, issue 2. In addition to Patriots, the train was also hauled by Claughtons and Royal Scots. This reminded me of Hornby's 00 gauge 'The Irish Mail Train Pack' (R2796M), which was a limited edition. The set included the rebuilt Royal Scot Class 4-6-0 No. 46127 OLD CONTEMPTIBLES in BR green livery.

There were three maroon BR Mark I coaches: composite passenger, buffet car and brake coach. It is disappointing that it did not include a Royal Mail sorting tender as this was the raison d'etre of the night mail! I understand that the pack is now a collector's item. How I wish that I had kept mine in pristine condition!

**BEST WISHES, GRAHAM RANKIN**

Hello Mr. Sikes

Further to my email of 21st August concerning a photo of Bangor station in issue 1, I've just noticed an error in issue 2. This is the photo on page 65 showing a 2-2-2 engine. This is CORNWALL and shows it in the Paint Shop.

However, it was a unique engine and certainly not a Dreadnought, these being a class of 2-2-2-0 Webb 3-cylinder compound engines.

**BEST WISHES, SIMON FOUNTAIN**

LMS 'Patriot' Class 4-6-0 No. 5521 RHYL at an unrecorded location in 1938 showing the nameplate without the crest.
PHOTO: COURTESY, THE LMS-PATRIOT PROJECT

Hi Pete,

I spent a relaxing time with a beer before dinner looking through both Midland Times and Eastern Times.

I think that they are excellent publications proving that the appetite for railway nostalgia shows no sign of abating. I like the combination of maps and plans as well as station and locomotive photos, and the mix of articles relating to specific events as well as locations, locomotives, and people history. The Transport Treasury lives up to its name.

I particularly enjoyed the article about Patriots named after seaside resorts in the North West, although you could argue that as Chair of The LMS-Patriot Project I should declare an interest. I am intrigued by the unusual combination of names sported by the Patriot class as a whole and indeed the process that led to their naming, which is well documented in John Jennison's RCTS book.

One of my favourite shots in the Midland Times is in the Stafford area article (page 13) where we see a "short" (indeed it is) train with a Palethorpes sausage van at rear. I picked up the existence of such beasts from the excellent shot you provided to me of Patriot No. 5507 ROYAL TANK CORPS at Carlisle with the same type of van in tow. This encapsulates the force of nostalgia, as all the ingredients – Patriots, fitted mixed traffic freight trains, the very concept of bulk conveyance of sausages by rail, and indeed the sausage company Palethorpes itself – are all long gone, although the brand names survived longer than anything else with a reincarnation in the 1990s.

In pedantry mode in the same article, I assume that the reference to Penkridge as a closed station (page 10), was meant to refer to the signal box, as I don't believe that the station itself was ever closed in its history. Perhaps someone with better local knowledge will prove me wrong.

I would like to subscribe properly to the Midland Times in future. It's all very well me having worked for BR on LNER, GNR and GER territory as well as LMS ground at Crewe, Chester and Stoke-on-Trent, but the combination of my father's LMS Engineer career background, his and my own Midland birthplace and (to top it all) a shared commitment to Derby County is what really pulls at the heartstrings.

COLIN HALL, THE LMS-PATRIOT PROJECT, CHAIR

*The 'Patriot' and Palethorpes van referred to above.*
PHOTO: © RAIL ARCHIVE STEPHENSON

Dear Peter,

Thank you for the kind words you say in your editorial (issue 2) about my ramblings on the LMS in Ireland. I think it's all there and hopefully correct!

I hope that I'm still being constructive when I make a comment on the picture at the top of Page 13 of the second issue?

The coach is not one of the ex-LNWR electric coaches, although it does have an air of them, being in the same style (pictured above). In fact there is no connection; it is a former Midland rail motor that was converted to be the personal saloon of the MR General Superintendent, Cecil Paget, at one time fitted with motor equipment and used with a motor fitted 'Spinner'(!) It was cascaded after the Grouping. It was MR number 2234 and LMS number 45010 in the saloon series. There is a link to your caption, in that it was built for use on the Lancaster/Morecambe/Heysham shuttles, which eventually were replaced by converted ex-LNWR stock in the early 1950s.

At some stage, I do not know when, it had been given to the use of the Signal & Telegraph Engineer at Crewe, remaining in such use until 1968. The control equipment was removed in 1927, so the transfer might date from then.

It was withdrawn in 1968, as it happens at the same time as George Dow retired as Stoke Divisional Manager from BR, who bought it and turned it into a holiday home. He gave it to the National Railway Museum where one would have thought it would be a prize exhibit. However, as is increasingly the case, the NRM decided to dispose of this precious relic and gave it to somewhere called 'The Chain Bridge Honey Farm' near Berwick. Whether it still exists I do not know, but there are some very concerning pictures of it looking to all intents and purposes abandoned at this location on the internet.

Presumably, your photograph shows it in use by the S&T Engineer's people on a job at Stafford that required sleeping accommodation.

YOURS, DAVID PEARSON, HAWORTH

---

An interesting cutting from a magazine/newspaper was sent to Midland Times by Kevin Robertson concerning arrangements made by London Midland Region officials, it reads thus:

*Train Stops for Lunch*

*The needs of travellers from Ireland, whose steamer docked 12 hours late at Stranraer on 25th October owing to heavy gales, were specially catered for by London Midland Region officials.*

*A special train was run to Euston, arrangements being made for a stop at Carlisle so that lunch could be taken in the refreshment rooms there, and a dining car was hurriedly brought from Manchester to join the train at Crewe.*

*As the train sped southward, the passengers were consulted as to their destinations, and the train was stopped at points most convenient to them.*

Totally at odds with the 'please don't travel' announcements that train operators make today!

SEND YOUR COMMENTS TO: MidlandTimes1884@gmail.com
OR WRITE TO:
Midland Times, Transport Treasury Publishing Ltd.,
16 Highworth Close, High Wycombe HP13 7PJ.

## 46225 DUCHESS OF GLOUCESTER
### An imposing image of one of Stanier's finest at Carlisle Upperby.
Photo: © Transport Treasury